Geographical Population Analysis:
Tools for the
Analysis of Biodive

METHODS IN ECOLOGY

Series Editors

J.H.LAWTON FRS
Imperial College at Silwood Park, Ascot, UK

G.E.LIKENS
The Institute of Ecosystem Studies, New York, USA

METHODS IN ECOLOGY

Geographical Population Analysis: Tools for the Analysis of Biodiversity

BRIAN A. MAURER
Department of Zoology, Brigham Young University
Provo, Utah 84602, USA

OXFORD

BLACKWELL SCIENTIFIC PUBLICATIONS

LONDON EDINBURGH BOSTON
MELBOURNE PARIS BERLIN VIENNA

© 1994 by
Blackwell Scientific Publications
Editorial Offices:
Osney Mead, Oxford OX2 0EL
25 John Street, London WC1N 2BL
23 Ainslie Place, Edinburgh EH3 6AJ
238 Main Street, Cambridge
   Massachusetts 02142, USA
54 University Street, Carlton
   Victoria 3053, Australia

Other Editorial Offices:
Librairie Arnette SA
1, rue de Lille
75007 Paris
France

Blackwell Wissenschafts-Verlag GmbH
Düsseldorfer Str. 38
D-10707 Berlin
Germany

Blackwell MZV
Feldgasse 13
A-1238 Wien
Austria

First published 1994

Set by Setrite Typesetters, Hong Kong
Printed and bound in Great Britain
at the Alden Press, Oxford

DISTRIBUTORS

Marston Book Services Ltd
PO Box 87
Oxford OX2 0DT
(*Orders*: Tel: 0865 791155
        Fax: 0865 791927
        Telex: 837515)

USA
Blackwell Scientific Publications, Inc.
238 Main Street
Cambridge, MA 02142
(*Orders*: Tel: 800 759-6102
            617 876-7000)

Canada
Oxford University Press
70 Wynford Drive
Don Mills
Ontario M3C 1J9
(*Orders*: Tel: 416 441-2941)

Australia
Blackwell Scientific Publications, Pty Ltd
54 University Street
Carlton, Victoria 3053
(*Orders*: Tel: 03 347-5552)

A catalogue record for this title
is available from both the British Library
and the Library of Congress

ISBN 0-632-03741-5

Contents

The Methods in Ecology Series

The explosion of new technologies has created the need for a set of concise and authoritative books to guide researchers through the wide range of methods and approaches that are available to ecologists. The aim of this series is to help graduate students and established scientists choose and employ a methodology suited to a particular problem. Each volume is not simply a recipe book, but takes a critical look at different approaches to the solution of a problem, whether in the laboratory or in the field, and whether involving the collection or the analysis of data.

Rather than reiterate established methods, authors have been encouraged to feature new technologies, often borrowed from other disciplines, that ecologists can apply to their work. Innovative techniques, properly used, can offer particularly exciting opportunities for the advancement of ecology.

Each book guides the reader through the range of methods available, letting ecologists know what they could, and could not, hope to learn by using particular methods or approaches. The underlying principles are discussed, as well as the assumptions made in using the methodology, and the potential pitfalls that could occur − the type of information usually passed on by word of mouth or learned by experience. The books also provide a source of reference to further detailed information in the literature. There can be no substitute for working in the laboratory of a real expert on a subject, but we envisage this Methods in Ecology Series as being the 'next best thing'. We hope that, by consulting these books, ecologists will learn what technologies and techniques are available, what their main advantages and disadvantages are, when and where not to use a particular method, and how to interpret the results.

Much is now expected of the science of ecology, as humankind struggles with a growing environmental crisis. Good methodology alone never solved any problem, but bad or inappropriate methodology can only make matters worse. Ecologists now have a powerful and rapidly growing set of methods and tools with which to confront fundamental problems of a theoretical and applied nature. We hope that this series will be a major contribution towards making these techniques known to a much wider audience.

John H. Lawton, Gene E. Likens

vi

Preface

We stand, I believe, on the brink of a scientific revolution in ecology. The revolution will not be brought about by a new conceptual paradigm, as occurred with Darwin or MacArthur, for example, but by new ways of measuring ecological systems. The sciences of genetics, systematics and many areas of medicine have been revolutionized by the micro-technologies of molecular biology. The new techniques of molecular biology have allowed new kinds of questions to be asked that could not even have been posed without a detailed knowledge of the molecular structure of the materials of inheritance. In the same way, new technologies for the measurement and analysis of ecological data are emerging that will allow the posing of new questions in ecology. Remote sensing technologies, including satellite imagery, provide the potential of measuring ecological systems at scales of resolution never dreamed of a few decades ago. Geographical information systems are becoming more sophisticated and provide researchers with the ability to manipulate and analyse huge amounts of data from ecological systems. Advances in the mathematical understanding of the nature of complex shapes and the dynamics of complex systems provide new tools for describing complex ecological phenomena. As these new techniques become integrated with the study of the living systems inhabiting our planet, I believe that ecology will be presented with whole new sets of phenomena to explain that will require new theories and paradigms.

One can already see the beginnings of such shifts in ecological thinking. Two decades ago, populations and communities were considered almost as if they were closed systems. Now, ecologists are aware of physical transport processes that affect population dynamics, landscape level processes that result from patterns in the interface between local 'patches' of different habitats, and regular patterns in statistical distributions of body size, abundance and geographical range areas in large collections of related species of many kinds of organisms such as beetles, birds, butterflies and trees. The realization that biogeochemical cycles link distantly located ecosystems together has led to an awareness of the global nature of ecological systems that is increasingly influencing the way ecologists view local ecological events.

When I first began to think about writing this book, my intent was to collect a number of loosely related statistical techniques that I and others had been working with for a number of years and to integrate them into a coherent approach to biogeography. As I made progress in compiling information on the techniques I discuss in this book, I began to realize that the common theme underlying these techniques was that they could all be used to describe all the populations of a species across its geographical ranges *as a single unit*. Current theory in biogeography, ecology and evolutionary biology is not sufficiently developed to say whether a geographical population is a single thing or a collection of things. Certainly there are precedents for considering geographical populations as single units rather than as simple collections of smaller units, but most ecological thinking and practice simply does not consider the possibility explicitly. With the advent of the technological revolution in remote sensing and geographical information systems, we are just beginning to be able to measure geographical populations as if they were single units.

This book aims to lay a groundwork for analysing geographical populations. The ultimate utility of the techniques will depend on whether general patterns emerge from the kinds of analyses I discuss. If so, such patterns will require explanations, and such explanations will require new theories. Although I do not claim to know the details of what these new theories will look like, I hope some of the examples that I present in this book will convince readers that there are some *very* interesting patterns out there waiting to be documented. I have had the privilege of seeing some of these patterns in the data sets that I have looked at, but I suspect that there are many more waiting to be discovered. More importantly, I believe such discoveries will lay the foundations of new theories about how the earth's ecosystems operate on large scales.

The techniques that I describe share one common characteristic: they are computer-intensive and require relatively sophisticated hardware and software to be successfully completed. Therefore, I have found it difficult, and in some cases impossible, to provide simple formulas and worked examples of the applications of these techniques. I have tried to present the results of some of the analyses we have carried out in my laboratory in a manner that emphasizes what they mean and how they might be interpreted.

In 50 years, I expect this book to have suffered one of two possible fates. The bleaker fate would occur if the kinds of analyses that I outline herein turn out to be relatively uninformative. If so, the book will be viewed as an eccentric collection of computer analyses that were only interesting because they could be done. The other fate, and the one of which I am more hopeful, is that the book will be outdated by advances in the technology and theory of geographical population analysis that would have pushed our understanding far beyond the humble beginnings offered in this volume.

Acknowledgements

A number of people have influenced me or helped me in various ways during the conception and writing of this work. Dr James H. Brown has had a considerable influence on the way that I think about ecological systems and geographical populations. Dr Marc-André Villard worked closely with me during much of the time I was preparing this book. He read and commented on several chapters and helped me develop my thinking on various points more clearly. Dr P. Legendre provided helpful comments on Chapter 4 and Dr K. Gaston on the manuscript. Greg Heywood and Nikkala Pack helped to translate my theoretical rantings into computer algorithms and code, a step that would have taken me much longer. Dani Montague and Leslie Clifford helped with the menial tasks associated with actually doing the statistical analyses. Research support was provided by the United States Environmental Protection Agency, Grant R81-8358-010, and by Brigham Young University. And most of all, thanks to Cathy Maurer for her love and patience; I can get pretty grumpy when I can't find the bug in one of my programs. Inspiration for the future was provided by Danny, John and Megan.

CHAPTER 1

Geographical population analysis and the conservation of biological diversity

In the minds of many biologists, we are experiencing an unprecedented decline in species diversity that has few, if any, parallels in the history of life on Earth. Perhaps only the profound declines in numbers of species observed at the Permian−Triassic boundary, when over 90% of all species of marine organisms became extinct, might compare to the present rate of extinction. The problem in making such comparisons, however, is that it is difficult to obtain estimates of current extinction rates. In part, this is because we must project extinction probabilities of extant species into the future in order to form an idea of exactly what course the current wave of extinctions will take. Such projections introduce a number of empirical problems for which appropriate methods of using available data bases have not been worked out in detail. These problems include description of spatial patterns of populations across continents, description of the degree of fragmentation of a population and establishing a relationship between population dynamics and the degree of population fragmentation.

The purpose of this book is to introduce some techniques for dealing with these problems. The approach that I will take is one of exposition and exploration rather than of deductive inference. This is primarily because we do not know enough yet about how populations work on large spatial scales to deal with many of the topics covered in this book in a mathematically rigorous fashion. Nevertheless, I believe that there is a sufficiently large body of data available that could be brought to bear upon the general problems related to current extinction dynamics if we had some techniques which would allow us to evaluate relationships between geographical distributions and abundance, and how these might influence the risk of extinction for a species.

Geographical populations of species are hierarchies of populations nested within one another. Local populations are nested within larger 'metapopulations'. A number of definitions of metapopulations have been given. Here I refer to the collection of local populations occupying patches in a landscape as a metapopulation. Metapopulations themselves are found within regional populations occupying specific types of biomes, such as tall grass prairie or shrubbsteppe. A species often occurs across several biomes, so that several regional populations are included within the same species. Sometimes these regional populations are sufficiently differentiated to be recognized as separate subspecies. This hierarchical structure of a geographical population makes it difficult to characterize and quantify. Yet in order to understand how changes in continental and global ecosystems might increase the threat of extinction for a species, it is necessary to be able to describe geographical populations in a quantitative manner.

Analysis of entire geographical populations, I believe, will turn out to be a very informative approach to conservation. The advantages of considering geographical populations stem from the increasing sophistication of techniques of remote sensing of biological systems and of computer systems capable of analysing geographical data. In most situations, remote sensing systems are currently unable to distinguish between species of plants and animals within most communities. However, it is not inconceivable that in the future, technological advances may pave the way for higher resolution of such systems. Ultimately, it may be possible to obtain for certain kinds of species population abundance estimates (or correlates) from remotely sensed data. In addition, on the ground efforts to census populations of plants and animals are likely to increase as governments become more aware of the need to monitor biodiversity. Currently the best of these census efforts are directed at bird populations, but other relatively easily censused species of animals and plants should be considered for similar survey efforts in the future. Such population data can be accumulated and analysed using large computer systems called Geographical Information Systems (GIS). The technology for GIS is rapidly evolving and becoming more affordable. This will make compiling and manipulating data on geographical populations more accessible in the future.

There are a few limitations that might be encountered in geographical population analysis. Species differ to varying degrees in the genetic differentiation of their local populations from one region of their geographical range to the next. Some species appear to be relatively uniform morphologically, and sometimes genetically as well. Others vary considerably across space. This suggests that sometimes what might appear to be a single species, may, on further analysis, turn out to be several well differentiated varieties. It might be profitable in some situations to consider distinct subspecies as separate geographical populations. Another problem might arise if within the geographical range of a relatively uniform species, different local populations experience profoundly different sets of ecological conditions. Such populations might show strikingly different patterns of population change in response to changes in continental ecosystems. It would be important to have some techniques available to analyse such variation in population responses within the geographical range of a species. Both of these potential problems can be addressed within the framework that I discuss in the following chapters. In fact, several of the techniques I will discuss explicitly consider variation in population abundance and dynamics across space, and these techniques might be profitably applied and modified to analyse the kinds of patterns that might arise from variation in either genetic composition or in environmental conditions across a species' geographical range.

Before entering into a discussion of techniques that might be useful in analysing geographical populations with an eye to addressing concerns in the looming diversity crisis, it is necessary to establish exactly what it is that we are attempting to analyse. In the following sections, I will examine some concepts of what biological diversity is and how it has been approached in the past. I will attempt to identify a particular set of questions that remain unanswered, yet are relevant to current thinking on the preservation of biological diversity. It is these questions that the reader should keep in mind as we explore methods relevant to the analysis of geographical populations that might tell us something about how diversity is being threatened by modern society.

1.1 Hierarchical nature of biological diversity

Most ecological textbooks define species diversity as the number of species found in a given area (e.g. Ehrlich & Roughgarden, 1987; Begon, Harper & Townsend, 1990; Ricklefs, 1990; Smith, 1990). This has traditionally been the basis for discussions of biological diversity. Although it is an empirically tractable exercise to count the number of species in a given area, such a procedure ignores some potentially important phenomena. A broader definition of biodiversity includes both the number of species and the extent of their genetic variability (see, for example, Office of Technology Assessment, 1987). The OTA definition explicitly includes systems of different sizes, ranging from the biochemical variation in DNA to the number of species in ecosystems (see also Norse *et al.*, 1986). Hence, in its broadest definition, biological diversity is the manifestation of virtually every known biological process. Studying all such processes is clearly a daunting task, and one that will not be attempted here.

1.1.1 Genetical, phylogenetical and ecological aspects of biological diversity

In order to circumscribe the phenomena that I wish to examine later in this book, it is necessary to examine the concept of biological diversity in its widest sense. One convenient way to do this is to organize our perceptions of biological phenomena into some type of hierarchical scheme. Hierarchical organization is clearly a characteristic of biological systems (Allen & Starr, 1982; Eldredge, 1985; Salthe, 1985; O'Neill *et al.*, 1986). Unfortunately, there are a number of relevant ways to create a hierarchical ordering scheme. Biological systems participate in various different sets of processes. Three of the most meaningful from the perspective of biodiversity are genetical processes, ecological processes and phylogenetical processes. Although each set of processes is related to others, commonly, different kinds of biologists often associate their careers with each of these different sets of processes, so for operational reasons, we can consider each of these sets of processes separately and see how they are related to different concepts of biodiversity.

Genetical phenomena are concerned with the storage, maintenance and transmission of information regarding biological structure. The most basic meaningful unit of genetic information is the codon, a triplet of three base pairs in an organism's DNA that codes for a specific protein. Codons are organized into genes, which contain information regarding the primary structure of a protein. Genes are in turn collected together on chromosomes in most organisms, and the complement of chromosomes of a cell comprises the cell's genome. Since all cells have the same set of chromosomes in an organism, the genome of a cell is identical to the genome of the entire organism. Populations of interbreeding organisms, each of which has its own genome, form demes. Demes are usually groups of organisms that are capable of interbreeding at the same point in space and time. Demes, in turn, form collections that are capable of sharing genes over longer periods of time as organisms move about from one generation to the next. These collections, according to the biological species concept, form relatively discrete groups that share genetic information with one another, but not with other groups (other species). From the genetical perspective, a species is a collection of genetic information, and the most interesting aspect of biological diversity is the amount of genetic information contained in, within and among a collection of species. Clearly, the description of the genetic hierarchy given here is a simplification, and many processes occur that could complicate the discreteness with which we have elaborated levels. What is important here, however, is that the diversity in a group of species is based on the amount of genetic information that the group contains.

Phylogenetical phenomena are closely related to genetical

phenomena, but emphasize the patterns of similarity by descent of different biological units. A gene on a particular chromosome may have originated by the duplication of another gene, followed by a series of mutation. The two genes would be related because they were both derived from a common ancestor (note that the common ancestor can be identical to one of the descendants). As genomes are replicated and transmitted through reproduction, it is possible to construct families of genomes that are related to one another by common descent. In turn, groups of families may form the majority of a deme, and demes can be related to one another by the fact that new demes can be founded by members of existing demes. If gene flow is sufficiently low, then some demes will share more recent common ancestors than others. Species can be thought of in a similar manner. When a new species arises from some speciation event (assuming that it is derived from a single species by fission), it will be more closely related to its parent species than to other species that do not share a common ancestor. Groups of species can be constructed in a hierarchical manner (and the hierarchy could extend below the species level) based on the criteria of common ancestry, so that each hierarchical group contains only those subgroups descended from a common ancestor. Such groups are called monophyletic groups, and monophyly is argued by many systematists to be the best criteria for constructing classifications (e.g. Wiley, 1981). Diversity in the phylogenetical sense is associated with the number of species found within monophyletic higher taxa (often referred to as 'clades') such as genera or families. As before, there are many biological complications that might modify the scheme outlined in this paragraph. Nevertheless, when biological diversity is examined in a phylogenetical context, the criteria for measuring diversity is quite different from that used for genetic diversity.

Ecological phenomena are defined by exchanges of matter and energy between biological units and their surroundings. The biochemical reactions carried on by proteins constructed from DNA require energy. That energy is derived from cellular stores of ATP. Individual cells, in order to produce ATP must transport glucose and other compounds across their membranes to provide fuel for ATP synthesis. To burn glucose, cells must take up molecular oxygen, and release carbon dioxide. Groups of cells form organ systems which specialize in different functions but which all require oxygen and release carbon dioxide as a tissue. Organisms are collections of organ systems that use oxygen to burn carbohydrates obtained either from internal production via photosynthesis or from the breakdown of ingested organic material. An organism uses some of the energy it obtains to maintain a certain level of physiological functioning, and whatever is left over is allocated to reproduction. As populations of organisms that can mate with one another utilize energy, the patterns of energy acquisition of the entire population will determine the vital rates of the population: birth, immigration, death and emigration. Patterns of population vital rates among species in the same clade will determine the rates of extinction, and to a certain degree, speciation as well. Species diversity in an ecosystem depends at least partially on the availability of energy (e.g. Turner *et al.*, 1987, 1988; see Wright *et al.*, 1991 for a review). While this view again oversimplifies to some degree the complexity of ecological hierarchies, the important point is that ecologists can view the same set of biological phenomena as the geneticist or the systematist, but from an energetic viewpoint. Consequently, the ecologist's criteria for describing species diversity and the processes that affect it are likely to be quite different from those of geneticists and systematists, although all three perspectives are valid.

1.1.2 Genetics, phylogenetics, ecology and conservation

The sciences of genetics, phylogenetics and ecology all have important information to contribute to the preservation of biological diversity. In this book, I will take the ecological perspective. But in taking this perspective, it is still important to realize that the other perspectives have much to contribute to an ecological understanding of diversity. In this section I will briefly consider how the genetical and phylogenetical perspectives contribute to the conservation of biological diversity.

Many biologists have become concerned that as populations of species are reduced and fragmented by human activities, they become genetically impoverished by the process of random fixation of alleles in small populations (Frankel & Soulé, 1980). This phenomenon is well known in theoretical population genetics. There are several consequences of this loss of genetic variation in small populations. Firstly, the likelihood of rare, deleterious, recessive mutations of alleles are more likely to be expressed, because in small populations, individuals tend to be more highly inbred (e.g. they share many alleles by common descent). This leads to inbreeding depression, where the fitness of individuals is decreased by being homozygous for recessive alleles at multiple loci that each have a detrimental effect on the individual. In the long run, the loss of genetic variation in a small population may lead to the inability of the population to respond adaptively to changing environments.

Although the genetic consequences of small population size may threaten a species with extinction in the long run, Lande (1988) argued that the demographical consequences of small populations pose a more immediate threat. It is likely that most historical extinctions that have resulted from human activities have been the result of the demographical vulnerability of small populations. Small populations are more likely to suffer drastic declines in abundance due to environmental and demographical accidents than are large populations, regardless of the genetic diversity that exists in the population. Lande argued that by the time a small population loses sufficient genetic variability to lead to inbreeding depression, it will already have had many chances to become extinct simply due to chance occurrences. Therefore, he argues that most management of endangered wild populations of plants and animals should focus on the ecological problems of small populations rather than on the genetic problems.

Phylogenetic information is critical in conservation because in order to assess species' losses, we first have to be able to count them. Calculations of extinction rates are completely dependent on the accuracy with which we can determine how many species are available to become extinct. Obtaining accurate estimates of the number of species has proved to be much more difficult than expected (May, 1988, 1990; Erwin, 1991; Gaston, 1991a,b). Clearly, there is a great need to train systematists who are capable of cataloguing natural diversity. But systematics goes beyond the simple listing of species, because it is necessary, in order to preserve diversity, that we understand the biological nature of species and how they are related to other species. These ideas are necessary in order to understand how new species originate and thus to understand how diversity is regulated. It is likely that the kinds of speciation mechanisms and their importance varies from taxon to taxon and from continent to continent (Otte & Endler, 1989; Brooks & McLennan, 1991).

1.2 Conservation of biological populations

The ecological perspective on conserving biological diversity is necessarily concerned with the demographical mechanisms that are responsible for population decline. The number of species of a particular taxon in a given region and their genetic diversity hinges on the ability of populations of each species to persist in the region, either by reproduction being sufficiently greater than death and emigration, or by the immigration of individuals into the populations of each species from elsewhere. Traditionally, population theory has been mainly concerned with populations that occupy continuous space. However, with the work of Taylor (1961, 1986), Skellam (1951) and others, it soon became evident that most populations were spatially discontinuous, and that this structure had profound consequences for understanding spatial and temporal patterns in populations (see, for example, Pulliam, 1988; papers in Gilpin & Hanski, 1991). Unfortunately, much of the theoretical work of the last several years has failed to come to a consensus regarding the relationship between populations' spatial patterns and their persistence. For example, Quinn and Hastings (1987, 1988) argued that increased fragmentation of a population might enhance persistence by spreading risks among loosely connected subpopulations, while Gilpin (1988) argued that increased fragmentation decreased the size of each individual subpopulation, making each more susceptible to stochastic population fluctuations that could lead to extinction. This question is of major significance to conservation biologists, since much of the habitat that is available for the conservation of species exists now, or will exist in the future in a patchwork of habitats rather than in large continuous blocks.

The remainder of this chapter will consider briefly the nature of spatially discontinuous populations and how this affects population dynamics, and especially population persistence. These ideas are necessary to motivate the context in which the methods developed in this book should be viewed. I will also present some definitions that will be used in the rest of the book, especially regarding patterns in geographical scale populations.

1.2.1 Spatial structure of populations

Populations form a nested hierarchy of spatial inclusiveness beginning with the territory or home range of the individual and ending with the geographical range of a species (e.g. Kolasa, 1989). Populations are spatially discontinuous at all levels. An individual does not use all of its territory or home range evenly, but tends to spend more time in certain regions than others. The successful completion of its life history depends on the amount and spatial pattern of resources or other requirements that are found within an individual's territory. Territories or home ranges may overlap, particularly in social species. Again, however, territories are not uniformly distributed in space, but tend to be clustered together in regions of suitable habitat. Often, territories in close proximity may differ significantly in their suitability, where suitability is defined as the abundance of requirements for successful completion of an individual's life history (i.e. survival and reproduction).

At yet a larger spatial scale, clusters of territories or home ranges may be found together in landscape patches that are more suitable than other patches. The intervening landscape matrix is often unsuitable for the species. Many times, if these spatially separated clusters of territories tend to undergo relatively independent dynamics, then they might function as separate populations. The collection of many of these partially independent populations is often referred to as a meta-population. The metapopulation is often thought of as a natural ecological unit that has its own set of dynamics derived in part from the dynamics of the populations that comprise it (which in turn are derived from the success of individuals in meeting their needs for reproduction and survival).

In a regional ecosystem or biome, there may often be more than one metapopulation. Since the geographical ranges of most species extend across the boundaries of more than one biome, some meta-populations may be distinct or semi-independent of others. Sometimes, these discontinuities have sufficient genetic consequences to warrant the naming of different collections of metapopulations as different subspecies. The geographical range of species as a whole, then, is also spatially discontinuous to varying degrees.

In this book, I will be dealing primarily with the spatial patterns of geographical ranges of species. The techniques developed in the rest of the book may be applicable to smaller scales of population spatial patterns, but the ultimate threat of extinction for a species hinges on the survival of a sufficient number of metapopulations to ensure that the species is capable of withstanding the inevitable changes in global and continental ecosystems that occur during periods of geological and climatic change. In addition, for some groups of species, most notably birds, comprehensive continent-wide censuses are available. Such information on geographical scale spatial patterns allows the development of quantitative methods for the description of spatial patterns.

One of the greatest concerns of conservationists has been the increased level of fragmentation of once relatively continuous populations into smaller units. As mentioned before, populations are fragmented on many scales, so the concern is not over fragmentation *per se*, but rather over the intensity of the fragmentation. By taking a relatively continuous, although certainly not homogeneous, set of similar habitats, such as deciduous forests, and breaking them up into small islands of forest surrounded by agricultural or urban habitats, the net effect is that of increasing the intensity of fragmentation, so that more of the original forested habitat is bordered by drastically different habitats in the fragmented landscape than in the original forested landscape. As we shall see, the discontinuous nature of populations makes obtaining objective, scale-independent measures of fragmentation by no means straightforward.

In describing the level of fragmentation of the geographical range of a species, it is necessary to realize that there are at least two different senses in which the geographical range of a species can be fragmented (Maurer & Heywood, 1993). The first sense deals with the perimeter of the geographical range. We can recognize that across a continent, there are some geographical regions that contain populations of the species and others that do not. The boundary between these two regions marks the edge of the species' geographical range. This boundary can be thought of in the same way as one can think of a coastline: an irregular boundary between two distinctively different geographical elements. In the case of the coastline, the elements are water and land. For a geographical range, the elements are habitat and nonhabitat. The shape of the boundary of a geographical range can be relatively smooth, with few isolated populations at the edge of the range, or it can be highly convoluted, with many small, isolated populations along its border. Intuitively, we would say that the second case, that of a geographical range with a highly convoluted boundary and many isolated populations, should be more fragmented than the first. In what follows, I will refer to this kind of fragmentation as 'areographical fragmentation', since it deals with the shape of the species' geographical range (Rapoport, 1982).

There is another sense in which a geographical range might be fragmented. A species might occur relatively continuously over space, but have very different densities in different locations. Heuristically, the population can be thought of as a mountain range, with density peaks and valleys. A species with a relatively smooth density topography would tend to show relatively gradual changes in population densities from one place to the next. On the other hand, a species with a jagged density topography would show very drastic changes in densities from one place to the next. This kind of fragmentation is called 'demographical fragmentation', since it deals with differences in local population dynamics across a species' geographical range.

Demographical and areographical fragmentation do not necessarily result from the same processes. Areographical fragmentation is determined by the presence of a population of a particular species. It reflects nothing about the quality of a given geographical region other than indicating where it is possible for the species to exist. In essence, it reflects only the extremes of habitat suitability for a species. Demographical fragmentation, by contrast, must reflect finer distinctions in the quality of the habitat for a species, since different regions of a species' geographical range differ in their ability to provide the elements necessary for successful completion of an individual's life history, and hence, for the stability, persistence and density of the population.

A species' population abundance is not constant across its geographical range, but varies in a regular pattern. Generally, a species' population density is highest near the centre of its geographical range, and declines towards the periphery (Hengeveld & Haeck, 1981, 1982; Brown, 1984; Hengeveld, 1990; Maurer & Villard, 1993). There is often a single peak of high density and relatively smooth changes towards the periphery. However, for a substantial number of species, there is more than one mode in population abundance. Nevertheless, even in those species with more than one mode, there is clearly a decline in population abundance towards the periphery of the geographical range. It is not clear, however, whether such patterns of spatial distribution of abundance show any regular changes as a species declines and approaches extinction. Without techniques to analyse these spatial patterns, and the degree to which they are due to fragmentation of habitats across a species' geographical range, it is not likely that we will be able to develop an understanding of the geographical indicators of extinction likelihood. The major goal of this book is to develop some of these techniques.

1.2.2 Spatial dynamics of populations

The major question brought up in the last section is 'to what extent are spatial patterns in geographical ranges indicators of demographical problems in a species?' This is clearly a major concern of conservation biologists, yet there has been relatively little progress made in discovering the answer. Most theoretical treatments of population vulnerability to extinction have centred around local breeding populations or metapopulations (Quinn & Hastings, 1987; Lande, 1988; Gilpin & Hanski, 1991). It is not likely that metapopulation dynamics are simply additive consequences of the dynamics of local populations, and similarly, it is not likely that geographical range dynamics are strictly reducible to the dynamics of metapopulations. Here, I briefly discuss some theoretical results for aggregated populations with the intention of establishing some of the basic theoretical questions that can be addressed with appropriate analyses of geographical range fragmentation.

Skellam (1951) was one of the first researchers to consider patterns of abundance across a species' geographical range. He used a model of diffusion across geographical space, where the diffusion constant represented the tendency of organisms to move about in geographical space. His model successfully predicted the rate of spread of muskrats across Europe after their accidental introduction in 1905. The limitation of his model, however, is its treatment of population diffusion as a continuous process. We have established above that populations are not continuously distributed through space, but are patchily distributed on several different scales simultaneously.

A slightly different approach was taken by Hengeveld and Haeck (1982). They considered diffusion models for the distribution of abundances across species' geographical ranges, but also considered a different, more biologically realistic model. They suggested that the general observation made by many ecologists was that ecological conditions worsened for a species towards the periphery of its geographical range. Therefore, they suggested that the distribution of abundance of a species across its geographical range represented a kind of optimal response surface, where the ecological conditions at a given location dictated an 'optimum' for that locality, and the distribution of these optima in space produced the characteristic distribution observed. Species declined in abundance away from regions of high abundance because the environmentally-determined optima decreased.

Brown (1984) provided a more detailed model for the geographical pattern of abundances seen in species. He suggested that ecological conditions in the environment tended to be autocorrelated in space, so that the decline in ecological conditions across geographical space for a species was relatively gradual as one proceeds from centres of population abundance. The spatial distribution of species' ecological requirements changes in space less rapidly than ecological conditions (due to gene flow), so the ability of species to use these requirements declines in space, and this leads to the observed decline in population abundance. Maurer and Brown (1989) developed this notion further by observing that the decline in ecological conditions over space should translate into declines in equilibrium population densities of local populations arrayed along a geographical gradient because ecological conditions determine the demographics of these populations. In populations near centres of abundance, population gains are sufficiently large in relation to population losses to ensure that these populations will have relatively high densities. Near range boundaries, population losses will approach population gains, and Allee effects may lead to lower population limits for population stability, so that these populations should be both lower in density and exhibit a greater tendency to become locally extinct when compared to more central populations.

The models of Hengeveld and Haeck (1982) and Brown (1984, Maurer & Brown, 1989) generally assume relatively continuous variation in population processes across space, a limitation they share with the earlier diffusion models of Skellam (1951) and others. But populations are not continuously distributed in space. As indicated above, they are patchily distributed on local and landscape scales. Does this patchy structure across the landscape preclude the use of demographical models of population abundance such as those suggested by Maurer and Brown (1989)? The answer to this question is exceedingly important, because such models have much to say about the persistence of populations over time and space, one of the great concerns of conservation biologists. Explicit tests of these models in spatially complicated population systems in space must await the development of statistical techniques to analyse the complexity of species' abundance distributions.

1.3 Geographical population analysis and conservation
The conservation crisis that seems to be happening all around us is clearly due to the widespread alteration of ecological systems across the planet. Yet ecologists have traditionally focused their attention on very local spatial scales and on very narrow temporal scales. Pimm (1991) argues that if we are to begin to understand these problems, we must begin to focus ecological investigations on larger spatial and temporal scales. In this chapter I have indicated that if we move to larger spatial scales, we inevitably must concern ourselves with describing the spatial complexity of species' populations. Conservation biologists have pointed out that biological populations are fragmented across landscapes, and we have indicated that this fragmentation occurs on several spatial scales. The contribution of ecology to managing these spatially complex populations is unclear at the moment because few ecologists have attempted to think about population dynamics on these larger scales.

This provides us with a good starting point for the methods to be discussed in this book. What I will attempt to do in the following chapters is to lay out methods for describing the complex spatial patterns of abundance of biological populations across their geographical ranges. I begin with a discussion in Chapter 2 about some recently developed techniques borrowed from geostatistics that seem particularly adaptable for use in geographical population analysis. Since I rely on many of these concepts in later chapters, I lay out some important ideas in this chapter that will be useful later. In Chapter 3 I describe methods for estimating the size, shape and orientation of geographical ranges. For relatively continuous distributions, these statistics can be related to the estimation of the major and minor axes of a bivariate probability distribution, where the probability is proportional to the population density of a species across its geographical range. These methods are not satisfactory for species with fragmented population distributions, so I discuss some analogues of size and shape using fractal geometry. Chapter 4 deals with aspects of the distribution of population abundance. The measures of fragmentation that are most relevant to conservation biologists are probably those that measure demographical fragmentation. It is likely that demographical fragmentation is more closely related to population decline than areographical fragmentation, since it deals with the rate of decline of populations across their geographical ranges and perhaps in time as well. Chapter 5 addresses methods for studying the temporal dynamics of geographical populations. Initially, I discuss models that assume the population is relatively continuous in space, so that local populations are tightly connected. I end the discussion by considering methods that can be used to study patterns of temporal variation in populations across geographical space, since these patterns are closely tied to the temporal persistence of species.

CHAPTER 2

Regionalized variable theory for geographical population analysis

There are a number of techniques from spatial statistics that could be applied to the analysis of patterns in geographical populations. In this book I focus mainly on a set of statistical techniques that have been pioneered by geostatisticians. These techniques are often referred to as regionalized variable theory (e.g. Matheron, 1963; Burrough, 1986). A regionalized variable refers to the outcome of a process that varies across geographical space. Such variation can be entirely random, in which case statistical description of the patterns will give relatively little information, or it may be at least partially deterministic. In the second case, the deterministic part of the process may be exceedingly complex, so that simultaneous description of all the causal factors determining the spatial pattern is not possible. In essence, what is needed in such cases is akin to statistical mechanics in physics. That is, even though specification of every detail of the process is not possible, it is still useful to find macroscopic descriptors of the results of the process that themselves may be amendable to statistical analysis. Ideally, such macroscopic descriptors should have properties that are sufficiently different for random spatial processes and for complexly organized ones so that the two can be separated.

The concept of a regionalized variable is particularly applicable to geographical patterns in population abundance because population dynamics have both random and deterministic components. The deterministic processes underlying population change in space and time are often complicated and varied for each species. Such processes include the abilities of individual organisms to use resources, the mode and magnitude of the dispersal capabilities of organisms, the patterns in abundance and variation of resources across space, and the temporal patterns of environmental change at various spatial scales (see Wiens, 1989 for further discussion of these kinds of ideas). Clearly, ecologists and biogeographers cannot hope to measure so many variables at the appropriate spatial and temporal scales to allow the construction and testing of appropriate hypotheses about geographical patterns in abundance. The goal of this book is to describe a series of techniques that can be used to analyse the macroscopic properties of geographical populations. To do this, I begin by describing some useful results from regionalized variable theory in the present chapter.

2.1 Regionalized variable theory

The distribution of population density across the geographical range of a species is determined by the spatial patterns of population change within local populations and by the diffusion of individuals moving among populations. Spatial population patterns are determined by a spatially variable process, that of local population change, and by processes that correlate the temporal dynamics of local populations. These processes can be separated into migratory processes, those that result in the movement of individuals from one local population to the next, and environmental processes that determine the spatial correlation of resources available to local populations.

Regionalized variable theory is a set of mathematical models that have been used extensively in the earth sciences to model spatial variation in physical processes across the earth's surface. These models, as with all mathematical descriptions of physical phenomena, are meant to approximate the results of spatial processes rather than to be causal descriptions of the mechanisms that underlie those processes. To be useful, such models assume that there are general similarities among different spatial processes that allow their representation using a relatively small set of general models. This use of a small number of general models to represent imprecisely a wide variety of processes forms the basic premise upon which modern statistical practice is built. In analysing biological populations, ecologists have traditionally sought to use mechanistically motivated models (the logistical and exponential models are the simplest) to develop theoretical statements that are tested rather haphazardly with data. Mechanistic models of spatial population dynamics are often mathematically complex and intractable, and their uses have been relatively limited (see Edelstein-Keshet, 1988 for a review). Regionalized variable theory provides a way to represent spatial population patterns without specifying a particular population model. The price for using such models, of course, is that questions about specific mechanisms cannot be posed with much precision. Given the little that we know about geographical populations of most species, it is unlikely that we have either sufficient data or insight to test hypotheses about specific mechanisms. The justification for using regionalized variable theory is that it provides a descriptive calculus for allowing the researcher to begin to study variation within and among species in the geographical patterns in their populations.

The purpose of this section is to introduce some basic concepts about regionalized variable models that will be used later in the book. Additional discussions of these topics can be found in Matherton (1963), Ripley (1981), Burrough (1986) and Haining (1990).

2.1.1 Preliminary considerations

Before describing the concepts from regionalized variable theory that I will use later in this book, it is necessary to discuss some basic definitions and concepts.

Any spatial pattern can be said to have two general properties that are useful to distinguish between. The first property is called the 'support' of the spatial pattern. The support refers to the total amount of geographical space occupied by the spatial pattern. Consider the collection of all sites within a specified geographical region. The support of a spatial pattern that occurs within that region is the set of all sites at which the processes that generate the spatial pattern occur. Attributes of the support that provide information about the processes generating the pattern include its size, shape and orientation. The second property of a spatial pattern is called its 'measure'. The measure of the pattern is a description of the variation in the intensity of the process that generates the pattern. The support of a geographical population is the set of all local habitats in which that species resides within a geographical region, such as a continent. The measure of that population is population density, which represents the intensity with which population processes operate within the geographical range of a species.

With the possible exception of a small number of rare and endangered species, there is no species for which we have sufficiently complete censuses to determine the properties of the support and measure of its geographical range. We are left with using sampling procedures to determine these properties. Sampling procedures also have certain properties that are useful to recognize (Fig. 2.1). The first has been referred to as the 'grain' of the sample. The grain is the smallest unit of space that can be distinguished by the sampling procedure. For example, if the samples being considered are 10 hectare (ha) census plots, then the grain of the sample would be 10 ha. Phenomena may be occurring within those census plots that might be important to the spatial pattern being studied, but they cannot be directly measured by the sampling scheme. They can only be inferred from the pattern. The second property of a spatial sample is its 'extent'. The extent of a spatial sample can be defined as the distance between the furthest points in the sample. The extent of the sample defines the maximum size of the pattern that can be detected by the sampling scheme. Processes may occur outside the extent of a sample that will influence the pattern obtained, but they will be difficult to infer from the pattern. In some cases, a temporal sequence of patterns measured in the same area at the same extent may be useful in inferring the workings of such a large scale process.

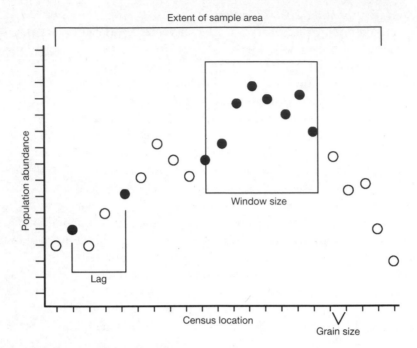

Fig. 2.1 Characteristics of spatial sampling schemes useful for geographical population analysis. Census locations are shown in a single dimension, as along a linear transect, but the most interesting sampling schemes for geographical populations are those that are two-dimensional, covering both latitudes and longitudes. For a two-dimensional sample, lag refers to the distance between sampling points, while the window size is the length of one side of a square that contains all the sampling points within the box. (Modified from Milne, 1991).

When analysing spatial data, two additional terms are useful in describing how the data are analysed. The first term is the 'lag'. When comparing sampled values of a spatial pattern, the distance between samples is called the lag distance, or lag (Fig. 2.1). The maximum lag is equal to the extent of the sample. Often, spatial processes can be inferred from changes in the behaviour of some statistical quantity calculated from measurements of the property as lag distances are varied. Such analyses can identify the spatial scales at which the process operates (see, for example, Milne, 1991). Another similar term that is useful in spatial data analysis is 'window size'. A window refers to a collection of sampled points, and its size is the distance between the furthest points in the sample. Windows can also be used to calculate statistical quantities that describe the measurements taken at each point in the spatial sample. As with lag distances, changes in these statistics with changes in window size can be used to infer the scale and nature of a spatial process.

2.1.2 The general regionalized variable model (GRVM)

Variation in a spatial process can be modelled using an additive model that assumes there is a deterministic and a stochastic component to the process. The stochastic component, in turn, is composed of a term that varies across space, and one that varies independent of space. Let the variable $X(i)$ be the value of a continuous random variable (i.e. X takes on values that are real numbers) at location i, then the value of $X(i)$ can be modelled as

$$X(i) = \mu(i) + \epsilon(i) \tag{2.1}$$

where $\mu(i)$ is a deterministic trend over space, $\epsilon(i)$ is a spatially dependent residual around the trend with a mean of zero, variance σ^2, and a spatial covariance (that is, values of this variable are correlated in space; see, for example, Burrough, 1986). The variable i in this model denotes continuous space, that is, the process varies continuously from one place to the next. We assume that the process is 'isotropic', which means that equation 2.1 defines the behaviour of the process regardless of the direction we move. We also assume that the process is 'stationary in its increments' (Haining, 1990). When this is true, the variance of the difference $X(i) - X(i + h)$, where h is the lag, depends only on h.

I will refer to the model in equation 2.1 as the general regionalized variable model (GRVM). The utility of the GRVM is that it explicitly includes spatial correlation among sites in the modelling of a spatial pattern. Therefore, if the process generating the pattern is spatially correlated in some manner, then samples from the process can be used to estimate the parameters of the GRVM. Information about the process can be obtained from these estimates of GRVM parameters. For spatial patterns in population abundance, the processes underlying population dynamics are thought to be spatially autocorrelated (e.g. the abundance of resources), so the GRVM seems to be an appropriate model to use to describe spatial variation in abundance.

To better understand the meaning of the model given in equation 2.1 consider the following statistical quantities that describe the model. Firstly, the average value of the random variable $X(i)$, or its expected value, is given by the deterministic part of equation 2.1, or

$$E[X(i)] = \mu(i)$$

where E denotes taking the expected value. The variance of $X(i)$ is given as

$$E[(X(i) - \mu(i))^2] = E[(\epsilon(i))^2]$$
$$= \sigma^2$$

The properties of the spatial pattern are studied by examining spatial covariance. The spatial covariance is a function of the lag. For most spatial processes, the spatial covariance should decrease with increasing lag distance. The covariance between $X(i)$ and $X(i+h)$ is the expected value of their product, which will be denoted by $cov[X(i), X(i+h)]$. This covariance is given by

$$cov[X(i), X(i + h)] = E[X(i) \cdot X(i + h)]$$
$$= cov[\epsilon(i), \epsilon(i + h)]$$

Therefore, the spatial covariance of the process is determined by the spatial covariance of the random term in equation 2.1.

An important quantity that is used frequently to analyse patterns of spatial covariance is the semivariance (see Rossi *et al.*, 1992 for a review of ecological applications). The semivariance can be called the variance of increments. It is obtained by calculating the variance of the difference between the random variables $X(i)$ and $X(i+h)$. This is given by

$$2\gamma(h) = E[(X(i) - X(i + h))^2]$$

which simplifies to

$$\gamma(h) = \frac{[\mu(i)]^2 + [\mu(i + h)]^2}{2} + \sigma^2 - cov[\epsilon(i), \epsilon(i + h)] \qquad (2.2)$$

In equation 2.2, the variance of $\epsilon(i)$, given by σ^2, is assumed to be constant over space. The first term in the equation represents the deterministic component of the semivariance and is referred to as the 'drift'. Often, the process being modelled is assumed to be stationary, so that the mean and variance do not change across space. In this case, the drift is zero. In practice, the drift is modelled by a parametric function, often a polynomial (see Burrough, 1986 pp. 160−161), and this function is subtracted from the $X(i)$s. The semivariance is then calculated from the residuals. Notice that the semivariance is negatively related to the covariance. This means that as spatial covariance declines towards zero, the semivariance will level off when the drift is zero. Notice also that the semivariance can be partitioned into the drift, which corresponds to the deterministic portion of the model in equation 2.1, and the variance and covariance terms for the random component of equation 2.1.

A 'semivariogram' is obtained when the $\gamma(h)$ is plotted against h (Fig. 2.2). The maximum value of the semivariance is called the 'sill', and corresponds to the value of the semivariance when there is no correlation between points at lag h. The lag at which observations are uncorrelated is often called the 'range' of the semivariogram. Values of the random variable $X(i)$ at points a greater distance apart than the range can be considered as statistically independent. When the process being modelled is stationary, then the sill represents the local variance of the random component of the model. In theory, $\gamma(0) = 0$, but in practice, the semivariogram has an intercept at zero lag that is greater than zero (Fig. 2.2). When $\gamma(0)$ equals a positive constant, this constant is called the 'nugget variance' (Haining, 1990). It represents the variability within sampling units. For example, for a collection of populations censused over a span of several years, the nugget variance would represent the year-to-year variability.

Equation 2.2 gives the theoretical semivariance. In practice, it is calculated from samples obtained through some spatial sampling procedure. Let x_i be an observation of a spatial pattern taken at location i, and let x_{i+h} be an observation of the same pattern taken at lag h, then an estimate of the semivariance at lag h is given as

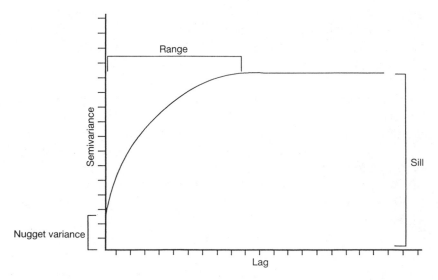

Fig. 2.2 Diagram of a hypothetical semivariogram.

$$g(h) = \frac{\sum\limits_{i=1}^{N(h)} (x_i - x_{i+h})^2}{2N(h)} \tag{2.3}$$

where $N(h)$ is the number of data points separated in space by a lag distance of h. The quantity $g(h)$ is an estimator of the semivariance given above.

2.1.3 Spatially discontinuous models

Some spatial processes do not vary smoothly across space, but are marked by distinct changes from one place to the next. The realized value at each point in space, i.e. the measure of the process, may itself be a continuous variable, and therefore be modelled by equation 2.1, but changes from one point in space to the next might not be smooth. Since most populations are discontinuous at multiple scales, as I discussed in Chapter 1, it is unlikely that population abundance will be smoothly continuous from one place to the next. In this section I consider a class of models that will prove particularly useful in analysing spatial variation in population abundance given the discontinuous nature of population structure.

Palmer (1988) simulated several different kinds of discontinuous spatial patterns that incorporated different kinds of effects at different scales. He found that the form of the semivariogram calculated from each of these patterns was related to the kinds of processes he incorporated into his simulations (Fig. 2.3). The pattern generated by a completely random process produced a flat line when log $g(h)$ was plotted against log h. For a pattern generated by a constrained random walk (called 'fractal Brownian motion' by Mandelbrot, 1983), the log−log plot was nearly linear (Fig. 2.3B). The simulated pattern that had a large scale trend but small scale randomness, the log−log semivariogram was flat for small lag distances, but increased linearly for longer lags (Fig. 2.3C). The opposite result was obtained for a pattern with small scale trends but large scale randomness: the log−log semivariogram increased for small lags but levelled off for large lags (Fig. 2.3D). Finally, a pattern was generated that was random at both large and small scales, but included intermediate scale trends, had a sloping log−log semivariogram at intermediate lags.

The lesson from Palmer's (1988) simulations is that the semivariogram contains information about the spatial structure of the pattern. Palmer included both stochastic and deterministic components in his simulations, but each component operated at different scales. From

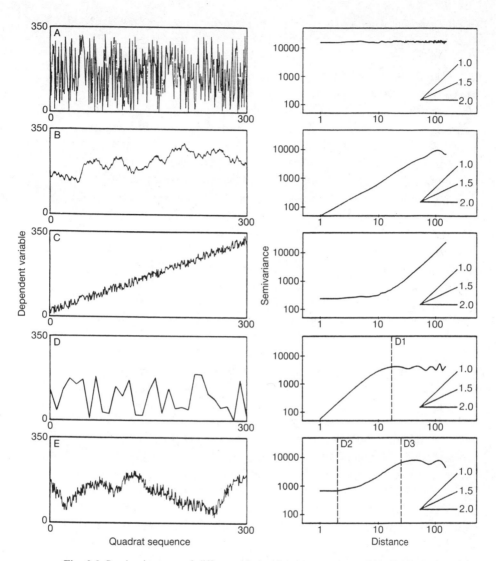

Fig. 2.3 Semivariograms of different kinds of patterns generated by simulations that incorporate processes operating at different scales. (From Palmer, 1988).

equation 2.2, recall that the semivariance included terms for both components. When the deterministic component of his model operated at small scales, the drift had a major contribution to the semivariance at small lags; when the deterministic component operated at large scales, then the drift contributed largely at large lags. In Chapter 4 I will discuss a technique using the semivariance to describe demographical fragmentation of geographical populations that incorporates many of Palmer's insights.

2.2 Regionalized variable models and geographical population analysis

The GRVM provides a method for describing spatial variation in patterns that are generated by processes operating at multiple spatial scales. Geographical populations are determined by a number of different processes operating from the scale of individual territories to the continental scale. One of the goals of the following chapters is to describe how the use of techniques related to regionalized variable models can be applied to geographical population data. Here I discuss briefly some general aspects of the relationship between geographical population analysis (GPA) and regionalized variable theory as I have described it above.

A distinction was made in the last chapter between areographical fragmentation and demographical fragmentation. In terms of regionalized variable theory, areographical fragmentation pertains to the support of a species' geographical range. Important properties of the support of a geographical range include its size, shape and orientation. Such quantities should reflect the range of environmental tolerances and resources that a species uses, because they deal primarily with the border of the geographical range. Demographical fragmentation deals with the measure of a species' geographical range, that is, the intensity of the population response to environmental conditions. This will reflect a number of ecological properties of individual organisms within species, including their efficiency in processing resources, their life history, and the kind of habitat that they select. These properties determine the population density of a species in local patches of suitable habitat. Spatial patterns in the abundance and quality of suitable habitat will lead to varying intensities of population responses and ultimately, to geographical patterns in population abundance.

Geographical population data is obtained from a variety of techniques which include intensive local surveys of individual territories to extensive broad censuses of populations across large landscapes. Each kind of survey implies a characteristic grain, and in theory should provide data on processes that operate only on scales greater than or equal to the grain of the survey technique. For example, extensive surveys that count organisms across large landscapes may not provide any detailed information on how organisms select habitat features, but may provide information on the range of ecological conditions within a landscape that a species is capable of using. The extent of a set of surveys will also be important in determining what questions the data obtained from them can be used to answer. Ideally, the extent of a set of surveys will include most or all of a species' geographical range. If so, then descriptions of the support and measure of that range can be obtained using the techniques described in this book. Some species will not be adequately surveyed, and descriptions can only be obtained for part of their range.

CHAPTER 3

Analysis of geographical range size, shape and orientation

In a forward-looking book, Rapoport (1982) called attention to the importance of studying the configuration of geographical ranges of collections of species. His approach was explicitly comparative, and assumed that even as species changed, the patterns exhibited by the collection of species did not, that is, species changed within boundaries set by ecological and evolutionary processes. He dubbed this approach 'areography', a name that has been adopted by others. Hengeveld (1990), for example, used the term areography to refer not only to the size, shape and orientation of geographical ranges, but also to the spatial pattern of population abundances across the geographical range and the dynamics of those patterns.

Areographical patterns have the potential to provide important information for conservation efforts. Although there is much that we do not know, many theoretical treatments imply that at least at the metapopulation scale, the size and shape of patches of suitable habitat has important consequences for the stability and persistence of populations on a landscape. However, on the continental scale, the relationship between shape and size of geographical ranges and continental population dynamics has received little attention (see Hengeveld, 1990 for a review). Clearly, if there are patterns apparent at the geographical scale that are correlated with the decline of the entire population of a species, then it would be important to identify what these characteristics were. Furthermore, the shape of a species' geographical range may indicate something about the extremity of conditions that individuals of the species are able to withstand (Root, 1988a), and this, in turn, may influence the likelihood of a species persisting in the face of changes in continental ecosystems.

The orientation of a species' geographical range usually corresponds to the topographical complexity of the continent and the orientation of major continental features, such as mountain ranges, deserts, rivers and large bodies of water (Rapoport, 1982; Brown & Maurer, 1989; Hengeveld, 1990). Such information is also important to conservation, because it indicates how a species responds to environmental extremes and rapid changes in ecological conditions.

Most analyses of using geographical range size have not differentiated between the amount of geographical space covered by the limits of a species' range and those areas within those limits actually used by the species. Gaston (1991c) referred to the first as the 'extent of occurrence' of a species and the second as its 'area of occupancy'. He pointed out that these two aspects of a species' geographical distribution may be very different, and many studies that have used extent of occurrence as a measure of geographical range size should have used area of occupancy, since the questions addressed were framed in terms of the species' ecological requirements. Clearly, area of occupancy is a better measure of a species' ecological limitations than is extent of occurrence. The difference between extent of occurrence and area of occupancy is important to keep in mind when using the techniques of analysis described in this chapter.

In this chapter, I present a variety of techniques for evaluating the size, shape and orientation of a species' geographical range. I start with a set of relatively straightforward techniques that assume that population abundance is distributed relatively continuously across the geographical range. The estimates of size, shape and orientation obtained using these are, of course, an approximation, since I have argued in Chapter 1 that geographical ranges are discontinuous at many scales. However, I believe that these techniques might be useful in analysing many data bases, such as those contained in distributional atlases, biogeographical monographs and similar publications. The limitation of these techniques is that they do not recognize the complexity of a species' geographical range boundaries. Therefore, I consider more realistic measures of range size and shape based on the assumption that geographical ranges belong to a class of geometrical objects called fractals. I will discuss fractals later in the chapter. Here I note that they are objects that can have very complex shapes and so are useful in representing many kinds of geographical objects such as coastlines and mountain ranges. The properties of fractals make them ideal models for description of areographical fragmentation.

3.1 Probability ellipse methods

Hengeveld's (1990; Hengeveld & Haeck, 1981) view of the distribution of population abundance across the geographical range as an optimal response surface suggests that it is possible to use abundance of the species at a local site to develop a measure of the likelihood, or probability, of a species being found within the region of the continent surrounding that site. Since not every part of a geographical range can usually be censused, the abundance of surrounding regions can be used to predict the probability that a species will be found in an unsampled region. The properties of this probability distribution can be used to derive descriptive statistics that contain information about geographical range size, shape and orientation. The general idea here is that we use the geographical centroid of a species' distribution, and the variances in abundance with latitude and longitude to measure the spatial extent of the distribution. Furthermore, the correlation between latitude and longitude will tell us something about the shape and orientation (i.e. the angle with respect to a latitudinal parallel) of the geographical range. It is important to note that measures of size based on latitude and longitude are biased in favour of species which occur at high latitudes because a 1° latitude−longitude block (latlon) is smaller (in km^2) at higher latitude. In what follows, transformed latitude and longitude values refer to transformation of latitude−longitude coordinates to distances (in km) from 0° latitude and longitude. This transformation corrects the bias introduced by measuring geographical range properties using latlons.

3.1.1 Description of methods

Suppose we divide the continent up into small boxes h km on a side. Let n_i be the number of individuals of that species found in the ith square on the continent. Let x_i be the transformed latitude and y_i be the transformed longitude of the box; then it is possible to calculate a series of statistics that describe the geographical range of a species. An estimate of the centroid of the geographical range, (x, y) is

$$\bar{x} = \frac{\sum\limits_{i=1}^{N} n_i\, x_i}{\sum\limits_{i=1}^{N} n_i}$$

$$\bar{y} = \frac{\sum\limits_{i=1}^{N} n_i\, y_i}{\sum\limits_{i=1}^{N} n_i} \tag{3.1a}$$

where N is the number of boxes where the abundance of the species is greater than zero. Estimates of the transformed latitudinal and longitudinal variances and covariance between them are

$$S_x^2 = \frac{\sum\limits_{i=1}^{N} n_i\, (x_i - \bar{x})^2}{\sum\limits_{i=1}^{N} n_i - 1} \tag{3.1b}$$

$$S_y^2 = \frac{\sum\limits_{i=1}^{N} n_i \, (y_i - \bar{y})^2}{\sum\limits_{i=1}^{N} n_i - 1} \qquad\qquad (3.1b)$$

$$S_{xy} = \frac{\sum\limits_{i=1}^{N} n_i \, (x_i - \bar{x})(y_i - \bar{y})}{\sum\limits_{i=1}^{N} n_i - 1}$$

If the presence or absence of a species in a square is the only information available, then use $n_i = 1$ if the species is present and $n_i = 0$ if it is absent. If some measure of abundance or rank abundance is available then let n_i be that measure. Thus, the n_i in these equations represent quantities used to weight latitude and longitude locations by the abundance of the species at that location. If that abundance is not known, then latitude and longitude locations are weighted equally.

The size and orientation of the geographical range can be character-ized by calculating an ellipse based on these statistics. If the distribution of population abundance forms a unimodal, bell-shaped distribution in space, then the statistics defined in equations 3.1 will approximate the parameters of a bivariate normal probability distribution. This is not likely to be the case for many species (Maurer & Villard, 1993). However, the ellipse calculated from these statistics can still be con-sidered to be an estimate of the relative size and orientation properties of the geographical range. We can then use them to make comparisons among different species.

We first must calculate the variance along the major and minor axes of the ellipse. The statistical techniques to do this have been available since Pearson (1901) and form the basis of the statistical technique known as principal components analysis. Here we use the geometrical description of the technique (see Pielou, 1984 for more details on the geometry of principal components analysis). The variances along the major and minor axes can be calculated by taking the eigenvalues of the variance−covariance matrix of the transformed lati-tudes and longitudes (calculated as in equation 3.1b). The variance− covariance matrix is

$$\Sigma = \begin{bmatrix} S_x^2 & S_{xy} \\ S_{xy} & S_y^2 \end{bmatrix}$$

The eigenvalues are given by

$$\lambda_1 = \frac{S_x^2 + S_y^2 + D}{2}$$

$$\lambda_2 = \frac{S_x^2 + S_y^2 - D}{2} \tag{3.2a}$$

where

$$D = \sqrt{(S_x^2 - S_y^2)^2 + 4S_{xy}^2} \tag{3.2b}$$

The eigenvalues can be used to construct an ellipse assuming that the size of the major and minor axes are 1 standard deviation in length. The area of this ellipse can be used as a relative measure of geographical range size. Species can be compared based on this area. A species with a large ellipse will occupy a relatively large area of the continent. By weighting the means, variances and covariance of transformed latitudes and longitudes by the probability of occurrence of the species in each of the squares, the resulting size of the ellipse is most heavily influenced by those geographical regions where the species is most likely to occur. Hence this measure of relative geographical range size will reflect the size of the area of occupancy of a species.

Measures of relative shape and orientation can also be obtained from the variances and covariance. Firstly, consider the eigenvalues of the variance−covariance matrix given in equations 3.2. The square root of each eigenvalue was used to construct the two axes of an ellipse. If the two eigenvalues are nearly equal, then the ellipse will be very round, and if they are equal, the ellipse will be a circle. Therefore, the difference between the two eigenvalues will be zero for a circle and will increase the more flattened the ellipse is. However, the difference between the eigenvalues will also depend on the size of the geographical range, so to make the measure of shape comparable among species, the difference should be divided by the sum of the eigenvalues. This measure of circularity, which I will call Γ, is given by

$$\Gamma = \frac{\lambda_1 - \lambda_2}{\lambda_1 + \lambda_2}$$

or, in terms of the variances and covariance

$$\Gamma = \frac{D}{S_x^2 + S_y^2} \tag{3.3}$$

where D is given in equation 3.2b. If the distribution is perfectly circular, then $\Gamma = 0$, and as one eigenvalue becomes much larger than the other, $\Gamma \to 1$.

The angle of orientation of the geographical range with respect to an arbitrary latitudinal parallel can also be calculated using the eigenvalues, variances and covariance. It can be shown that the following trigonometrical identity holds (Pielou, 1977 p. 335)

$$S_x^2 \cos \theta + S_{xy} \sin \theta = \lambda_1 \cos \theta$$

where θ is the angle of the major axis of the probability ellipse with respect to latitude. Solving for θ gives

$$\theta = \arctan \left(\frac{\lambda_1 - S_x^2}{S_{xy}} \right) \qquad (3.4)$$

where $-90° < \theta < 90°$.

The statistics described in the preceding paragraphs allow the calculation of relative measures of size, shape and orientation of a species' geographical range from a sample of locations or abundances of the species across a continent. A number of such data sets exist. Floral and faunal atlases of various sorts divide a continent up into geographical blocks and give the presence and absence or abundance of each species in each block. Such data, as coarse as they may be, provide basic geographical information that can be analysed using the methods developed in this section. Perhaps one of the best faunal surveys conducted for an entire continent is the North American Breeding Bird Survey (BBS) conducted annually by the United States Fish and Wildlife Service and the Canadian Wildlife Service. Because I will use this data base to illustrate techniques throughout this book, I will discuss it in the next section. Examples of the statistics described above calculated using BBS data are described in section 3.1.3.

3.1.2 The Breeding Bird Survey data set

The BBS is a large data base on the population abundance of birds breeding on the North American continent (Robbins, Bystrak & Geissler, 1986). The first surveys were initiated in the early 1960s, and by the end of that decade surveys were being conducted across the North American continent. Figure 3.1 shows the current distribution of BBS routes across North America. Notice that there are some limitations of the data base evident from Fig. 3.1. Firstly, the geographical distribution of routes is uneven, with the highest density of routes occurring in the eastern United States. BBS personnel have made continued attempts to increase the intensity of sampling in the western regions

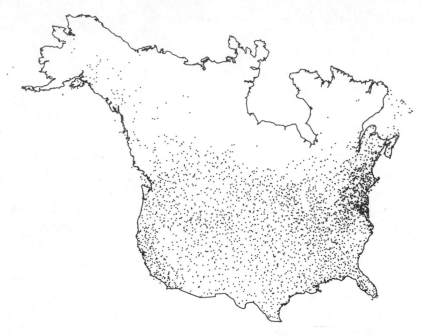

Fig. 3.1 Distribution of the over 3000 BBS routes censused at least once between 1965 and 1989. Notice that there is a lack of coverage at high latitudes across the continent. These locations define the extent of the geographical sample represented by the BBS. In the analyses that are discussed in this book, data analysis procedures were carried out as if there were a boundary along the region of no coverage at high latitudes. Therefore, no analyses were extrapolated beyond the extent of the data set. The minimum possible grain of the data is defined by a window size of 40×40 km (1600 km^2), since that is the length of a BBS route. In reality, the grain of the data set is much coarser, since most census routes are located at distances greater than 40 km from the nearest route and not all routes are censused every year.

of North America, but there is still a problem with undersampling. Secondly, the northern region of North America, particularly in the Northwest Territories, is virtually uncensused. Problems of access are probably responsible for this lack of coverage. Finally, the BBS ends at the border between the United States and Mexico, an obviously arbitrary border with respect to breeding birds.

Individual BBS routes are run as follows. An observer drives a 40 km route stopping every 0.8 km. At each stop, the observer counts all individual birds of each species seen or heard within a 0.4 km radius for 3 minutes. This is repeated 50 times. Thus, the basic datum is the number of times a species is seen on a 40 km route. The data are usually recorded by 'page', where each page contains counts for one-fifth of the route, so theoretically, the resolution of the BBS is finer than the 40 km route. However, this level of information is seldom used, and in subsequent analyses I will not deal with this aspect of the BBS.

The spatial scale of the observations taken limit its usefulness for asking certain kinds of questions and may bias the technique against locating certain kinds of birds. Only habitats that happen to be located near roads will be censused. If roads are randomly located with respect to habitats, this may not lead to a serious bias, but the problem will be most evident in western North America, where some habitats are not accessible. Birds that are regularly common in such habitats may therefore be under-represented in the BBS censuses. For this reason, there is some concern that analyses that compare different species may be inappropriate. Thus, as with all data analyses, caution must be used in interpreting the results. Results of many other studies using very different kinds of data sets, however, have confirmed that patterns obtained using the BBS data are not unique to that data set (see, for example, Strayer, 1986; Brown & Maurer, 1987, 1989; Maurer & Brown, 1988; Morse, Stork & Lawton, 1988; Ford, 1990; Maurer, Ford & Rapoport, 1991). Although the common patterns seen among these different data sets could potentially be due to very different kinds of biases, the similarities observed among data sets suggest that patterns obtained from the BBS show up in other data sets, often collected with very different kinds of methods. Some of the patterns seen in the BBS, however, have not shown up in every data set examined (see, for example, Lawton, 1989). For the present purposes, I will assume that the BBS provides an adequate, though potentially biased, representation of the abundances of most species across the North American continent.

3.1.3 Example analyses using the BBS

Geographical range centroids, variances and covariances were calculated from transformed latitudes and longitudes for about 390 species of North American terrestrial landbirds for which the BBS gave a sufficient sample size as follows. The latitude and longitude coordinates for each BBS census route were transformed into km from 0° latitude and 0° longitude, respectively. The number of individuals of each species was averaged over the years that the routes were run, and used as the n_i in equations 3.1. These estimates were then used to calculate estimates of the relative size, shape and orientation of the geographical ranges of species using the methods outlined in section 3.1.1.

Using the eigenvalues of the variance−covariance for transformed latitudes and longitudes, an ellipse was formed for each species by taking the length of the major axis as $2\sqrt{\lambda_1}$, or 2 times the standard deviation of the first principal component. The length of the minor axis of the ellipse was taken as $2\sqrt{\lambda_2}$. The area of this ellipse is an estimate of the relative size of the area of occupancy for each species. Gaston (1991c) suggested that area of occupancy be correlated with extent of occurrence for groups of species to assess the relationship between these two measures of geographical range size. Brown and Maurer (1987) calculated the size of a species' geographical range by planimetry of range maps contained in a field guide to North American birds. I used their estimates of range size as the extent of occurrence in order to correlate the two measures of geographical range size for North American terrestrial landbirds. Only species with a substantial portion of their geographical range contained in North America were included in the analyses described below. There were about 50 species used in these analyses that had ranges extending into Mexico that may have had estimates of their area of occupancy under-estimated by the BBS data set, although nearly half of these did not have enough data on the BBS to be included.

For species of North American landbirds, there is a strong positive correlation between the relative area of occupancy, measured as the size of the ellipse constructed as described in the previous section, and extent of occurrence (Fig. 3.2A). Examination of the residuals around the regression line in Fig. 3.2A indicated that there were no characteristics that uniquely identified species whose area of occupancy was poorly represented by its extent of occurrence. Neither area of occupancy (Fig. 3.2B) nor body mass (Fig. 3.2C) were correlated with the residuals, although the variance among residuals did seem to increase slightly with decreasing extent of occurrence (Fig. 3.2B). An analysis of variance of the residuals among the five trophic groups described by Brown and Maurer (1987) indicated no significant differences among trophic groups ($F = 1.63$; d.f. $= 4$, 320; $P = 0.17$). Individual species that had excessively large residuals showed no distinctive characteristics (Table 3.1). For some of these species, the area of occupancy estimate may have been biased by sampling problems. For example, three species of wood warblers, Lucy's Warbler (*Vermivora luciae*), Painted Redstart (*Myioborus pictus*) and Red-faced Warbler (*Cardellina rubifrons*) have small geographical ranges that extend into Mexico, where there are no BBS routes. These species represent a small fraction of the total used in these analyses. Considering all species, however, the correlation is strikingly good, and I conclude that area of occupancy and extent of occurrence are positively related for North American landbirds.

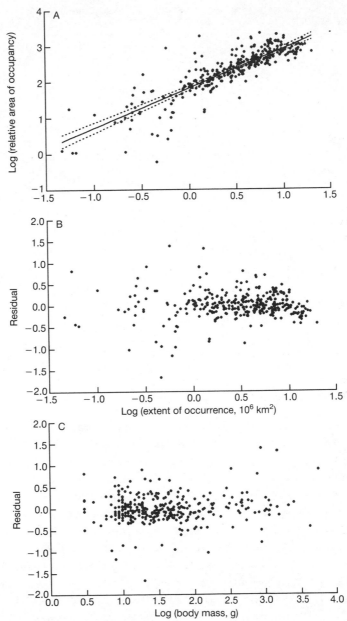

Fig. 3.2 (A) Relationship between the relative area of occupancy calculated as the logarithm of the area of an ellipse obtained from weighted BBS data and the extent of occurrence described as the area of the North American continent included within the species' geographical range boundary. The R^2 is 0.74, and the regression is significant. The dashed lines are 99% confidence intervals about the line. See text for details of the calculations. (B) Residuals from the regression line in (A) plotted against area of extent. Note the absence of trends in the residuals. (C) Residuals from the regression line in (A) plotted against the logarithm of body size. Note the absence of trends in the residuals.

Table 3.1 Species with residuals >2 standard deviations away from the regression line shown in Fig. 3.2

Species	Geographical range size ($10^6\,km^2$)	Trophic guild	Body size (g)	Bias
Meleagris gallapavo	1.18	Herbivore	5808	+
Leptotila verreuxi	0.21	Herbivore	153	−
Elanus caeruleus	0.33	Carnivore	332	+
Parabuteo unicinctus	1.47	Carnivore	885	−
Buteo albicaudatus	0.58	Carnivore	883	+
Falco peregrinus	3.40	Carnivore	781	+
Pandion haliaetus	1.32	Carnivore	1486	+
Nyctidromus albicollis	0.47	Insectivore	53	−
Selasphorus sasin	0.06	Nectarivore	3	+
Myiarchus tuberculifer	0.46	Insectivore	20	−
Ammodramus caudacutus	1.18	Granivore−insectivore	19	+
Cardinalis sinuatus	1.45	Granivore−insectivore	35	−
Vermivora pinus	1.40	Insectivore	8	+
V. luciae	0.65	Insectivore	7	−
Myioborus pictus	0.65	Insectivore	8	−
Cardellina rubrifrons	0.41	Insectivore	10	−
Anthus spinoletta	3.79	Insectivore	21	+

+, estimate of relative area of occupancy too large; −, estimate of relative area of occupancy too small.

Latitudinal and longitudinal centroids for geographical ranges of North American landbirds calculated from the BBS data set indicated a unimodal pattern for latitudinal range centres but a multimodal pattern for longitudinal range centres (Fig. 3.3). Peaks in the longitudinal range centres corresponded to prominent geographical features such as mountain chains. The eastern-most peak in longitudinal range centres corresponded to the location of the Appalachian Mountains, the next to the Great Plains, the third to the Rocky Mountains and the last to the Sierra Nevada−Cascade Mountains. Range shapes calculated from transformed latitude−longitude values for most species were relatively flattened ($\Gamma \simeq 1$), but were strongly skewed towards circular shapes (Fig. 3.4). The statistical distribution of angles of orientation of geographical ranges was highly peaked around zero, indicating that most species' geographical ranges were longest in an east−west direction, with a few ranges oriented in a more north−south direction (Fig. 3.3).

The patterns of variation in the size, shape and orientation of geographical ranges of species across the North American continent

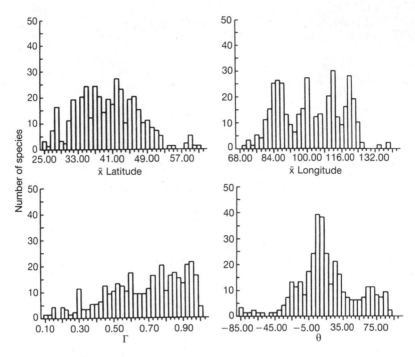

Fig. 3.3 Distributions of statistics describing the location, shape and orientation of geographical ranges of over 300 species of North American birds.

showed some interesting correlations between the characteristics of ranges and the location of their centroids. To look at the patterns of variation, I divided North America into 16 blocks of roughly similar size, and averaged the size, shape and orientation properties of all species that had their range centroids contained within the block. There was an increase in the size of ranges located further north (Fig. 3.4). This is most likely due to the fact that North America is shaped roughly like an inverted trapezoid, with more area located at higher latitudes. Ranges located in the centre of the continent tended to be more circular in shape, those located along the periphery of the continent were more flattened. Along the Atlantic and Pacific coasts, geographical ranges tended to be oriented so as to reflect the presence of mountain ranges, the Appalachians in the east and the Rockies in the west.

Although the analyses presented here should be considered preliminary, they indicate that it is possible to obtain biologically meaningful estimates of geographical range size, shape and orientation from extensive survey data. Few data sets are available that contain infor-

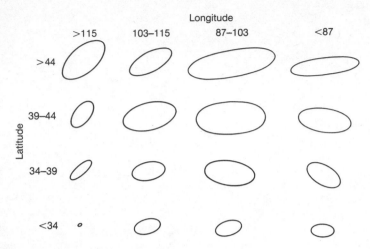

Fig. 3.4 Schematic diagram showing the relative sizes, shapes, and orientations of the geographical ranges of species with range centres (weighted by abundance) falling within different portions of the North American continent. Notice the increase in range size with increasing latitude.

mation on variation in abundances across a significant portion of a continent, as is the case for North American birds. However, the techniques described here can be used with presence−absence data as well, and may prove useful for groups of organisms where relatively detailed data are available on the spatial patterns of presence and absence.

3.2 Areographical fragmentation

Consider for a moment the problem of drawing a boundary on the geographical range of a species. In most field guides, the edge of the geographical range is drawn as a distinct boundary, outside which the species is rarely seen. The important concept here is 'rarely'. Clearly, individuals of the species may occasionally colonize or wander into geographical regions where the species is not usually seen. But from year to year, the boundaries of a species' range may fluctuate so that some years a species might be fairly common within a region and other years the species might be gone altogether.

There is another not so obvious difficulty, however. In any given year, the boundary between where the species is found and is not found depends on the scale with which we measure the boundary. If we use a map with a large scale, say $1:1\,000\,000$, then we may not be able to recognize whether the species is present in large blocks of land. If we drew an outline on this map of regions where the species was apparently present and where it was not present, then we would have an estimate of the species' geographical range boundary. Now suppose we looked at the same boundary on a series of maps at a finer scale, say about 1 : 250 000. On these maps, we may notice regions that were classified as containing populations of the species on the larger scale map, but on the finer scale map are obviously places where the species is not found. Similarly, other regions on the fine scale map where the species obviously occurs may have been classified as not containing the species on the coarser scale map. The reason for these misclassifications would be because the first map at the coarser scale had sufficiently poor resolution that it was impossible to differentiate between regions that contained the species and those that did not when those regions were below a certain size.

This phenomenon can be better understood by a hypothetical example. Suppose that we were interested in drawing a range boundary for a bird that breeds in forests in the eastern United States. Assume the species cannot breed anywhere but in closed canopy forest. As we approached the western boundary of the species, we would find it sporadically occurring in small woods and along forested streams. If we plotted the range boundary on a map of the United States, we would most likely draw a continuous line along the western boundary. However, suppose that boundary went through Indiana. On a map of Indiana, we might be able to identify unforested regions of the state where the species could not possibly occur, that on the large map of the United States appeared to be part of a continuous boundary. If we looked at the map of an individual county within Indiana, we might further see that forested regions were really broken up into small woods. Regions between small woods that appeared to be relatively continuous forest on the map of Indiana were actually discontinuous patches of forest, so that some places that would have been classified as habitat on the Indiana map, were in fact not suitable for the species. Thus, the range boundary that appeared to be continuous on larger scale maps was in reality discontinuous. The point is that even if we recognized the discontinuous nature of the range boundary at the small scale, we could not represent that structure very well on a map of sufficiently large scale.

The argument in the preceding paragraph could be repeated for finer and finer scale maps, until we got down to the scale of individual home ranges or territories. Clearly, the scale dependency of our ability to recognize geographical range boundaries will render estimates of geographical range size, shape and orientation, as calculated in the previous section, dependent on scale as well. This limits our ability to make comparisons among species to those sets of species for which we are confident our measures occur at the same scale. But more importantly, we are left with no objective measures of geographical range boundaries, and consequently, on the level of areographical fragmentation of the range. One solution to this problem is found by the use of the geometrical concept of fractals. In the next section, I discuss how the use of fractal geometry allows us to develop scale invariant measures of areographical fragmentation.

3.2.1 A brief introduction to fractals

In the rest of the book, I will use the concept of a fractal dimension often, so it is important to develop some understanding of what it means. In this section I briefly discuss the concept. Much more extensive treatments are given by Mandelbrot (1983) and Peitgen and Saupe (1988). Sugihara and May (1990) discuss some of the applications of fractals to ecological problems. More technical discussions are given by Barnsley (1988), Falconer (1990) and Schroeder (1991).

Consider the objects encountered in Euclidean geometry. Each kind of object has a dimension. A point has a dimension of zero, a line has a dimension of one, a circle or square is two-dimensional, and so forth. These objects do not correspond to anything in the real world, because the objects that comprise the world do not have smooth, continuous surfaces and edges. Instead, objects such as coastlines, mountain ranges and geographical ranges of species are jagged and discontinuous. As I argued in the previous section, the boundary of a species' geographical range will depend entirely on the size of the map that is used to draw it. Likewise, the length of a coastline will depend on the size of the ruler used to measure it. The smaller the ruler, the longer the coastline. In fact, there is a regular relationship between coastline length and ruler size (Mandelbrot, 1983). Let h be the size of the ruler used to measure the coastline, then the length of the coastline, $L(h)$, is

$$L(h) = a \left(\frac{1}{h}\right)^D \tag{3.5}$$

where a is a constant. The number D is called the fractal dimension of the coastline. If the coastline is a perfect straight line, then $D = 1$ and the length of the coastline will be a constant, independent of ruler size. If the coastline is jagged, then $D > 1$ and the measured length of the coastline will increase as ruler size gets smaller. Coastline length is scale dependent, but the fractal dimension of the coastline does not depend on the ruler size, and is therefore 'scale invariant'. When we compare different geographical ranges, it would be helpful if the properties of the geographical range used in the comparisons possessed this quality of scale invariance. The measures of fragmentation proposed in this and the next chapter are scale invariant, at least over a reasonable range of measurement scales.

What does the fractal dimension of a curve like the coastline tell us? The more broken-up or jagged the coastline, the higher the fractal dimension. Smooth, continuously varying curves have low fractal dimensions; highly broken-up, discontinuous curves have high fractal dimensions. This pattern extends to higher dimensions. Consider the three patterns in Fig. 3.5. To obtain these patterns, Mandelbrot (1983) ran three simulations, each of which was done with an increasingly high fractal dimension. Although Mandelbrot does not discuss the details of his simulation technique, the general idea for generating such surfaces is outlined by Peitgen, Jürgens and Saupe (1992). In the first stage of the simulation, a large square is constructed, and a height is chosen at each corner of the square by selecting a number from a Gaussian (normal) distribution. A point is located in the centre of the square, and its elevation is obtained by interpolating from the four heights of each corner and adding an appropriate number from a Gaussian distribution with its variance reduced by a factor of R, where

$$R = \sqrt{\frac{1}{2^H}}$$

The quantity H is related to the fractal dimension of the resulting pattern. Then, the original square is divided into four smaller squares, each of the same size, and the additional four points corresponding to common corners that do not have heights assigned are also interpolated, with increments again added that are drawn from a Gaussian distribution with its variance again reduced by a factor of R. This process is continued, so that a rugged surface is obtained on a grid. An arbitrary contour on this rugged surface (usually zero) is used to obtain an outline of the 'coastline' of the surface. It is these outlines that are shown in Fig. 3.5. Each was obtained using a different value for the parameter H. The differences in the degree of fragmentation among the patterns are evident. If one considers the three patterns to be archipelagos, than the lower archipelago, though it has approximately the same area as the others, has that area broken into much smaller pieces.

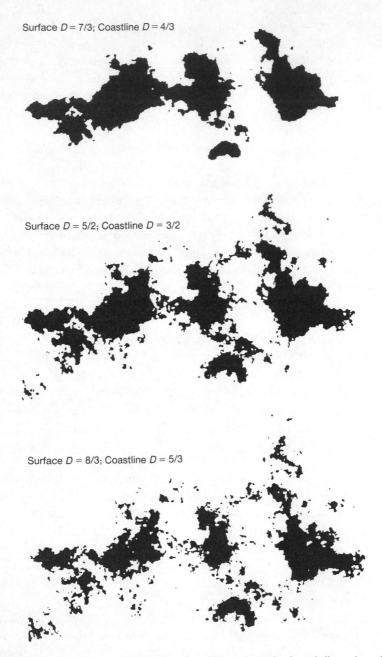

Surface $D = 7/3$; Coastline $D = 4/3$

Surface $D = 5/2$; Coastline $D = 3/2$

Surface $D = 8/3$; Coastline $D = 5/3$

Fig. 3.5 Three simulated sets of patches that differ only in the fractal dimension of the mathematical process used by Mandelbrot (1983) to generate them. The spatial pattern of fragmentation is related to the fractal dimension, suggesting that in general, fractal dimensions can be used to describe spatial patterns such as geographical ranges.

The patterns in Fig. 3.5 are similar to what we might expect to see if the border in each pattern were taken to be the outline of a species' geographical range. Measured as such, we would actually be considering the extent of occurrence of a species rather than its area of occupancy. The fractal dimension of the extent of occurrence of a species can be taken as a measure of areographical fragmentation. The degree of complexity of the border between where a species is found and where it is not found must reflect some of the fundamental constraints on populations of the species near the edge of its geographical range.

3.2.2 Measuring areographical fragmentation

There are a number of ways to compute fractal dimensions of objects. I will discuss two methods in this section that are relevant to areographical fragmentation. The first is called the 'box dimension'. To calculate the box dimension, the minimum number of square windows, or boxes, with sides of size h needed to cover the outline of a species' geographical range is counted (Williamson & Lawton, 1991). This number, call it $N(h)$, is obtained for several different box sizes. A simple linear regression is then calculated using least squares

$$\log N(h) = \beta_0 + \beta_1 \log (1/h) \tag{3.6}$$

The estimate of β_1 obtained from the regression is an estimate of D_B, the box dimension (Barnsley, 1988; Voss, 1988; Peitgen, Jürgens & Saupe, 1992). It has also been referred to as the capacity dimension in the literature on nonlinear dynamics (see, for example, Parker and Chua, 1989). In later chapters, we will encounter other ways to calculate fractal dimensions that have different interpretations. The box dimension of a geographical range is a descriptive statistic that can be used as a measure of areographical fragmentation. The box dimension of a pattern in two-dimensional space approaches two the more of that space the pattern covers. A pattern with many holes and peninsulas will have a lower box dimension than a pattern that is nearly solid and has a smooth boundary (Milne, 1991). Thus, the box dimension is a measure of the ability of the spatial pattern to fill space.

In practice, the measurement of the box dimension requires one to be aware of a few points prior to doing the regression. For geographical ranges, calculating the box dimension by hand is possible, but from personal experience, I recommend it only to acquaint oneself with the algorithm. It is much faster, and probably more objective, to let a computer count boxes. Therefore, it is necessary to have a computerized representation of the borders of the geographical range of each species for which a fractal dimension is desired. Such a representation can be obtained by digitizing the image of a map that outlines a species' geographical range or by accumulating estimates of spatial locations where a species' local abundance drops to zero. The second procedure requires that spatially extensive samples of abundance for the species are available, such as those available from the BBS (see section 3.1.2). Objective methods for estimating where the species' abundance drops to zero are discussed in the next chapter.

It is important to point out here that the size of the smallest box used to calculate the box dimension cannot be smaller than the resolution of the map. There are at least two reasons for this. Firstly, the resolution of the map is usually based on the resolution of the data, and using a box size smaller than the resolution may require one to extrapolate beyond the data. Secondly, even if the resolution of the data is smaller than that of the map, the processes generating the geographical range boundary may change as one looks at finer scales. To see this, recall that geographical ranges are discontinuous on many scales. Locally, the presence of a species may depend on the presence and abundance of a specific kind of resource. Along a geographical gradient, however, many different resources may come into play in specific populations to limit a species. At the geographical scale, the boundary indicates only that some resource, or combination of resources, is not available to individuals. The particular kind of resource is unimportant, and need not be the same in different parts of the range. Hence, if the box size used is too small, the analysis will change from one reflecting the general tendency of a species to be limited across its geographical range, to one reflecting multiple instances of specific kinds of resources limiting the species in different parts of its geographical range.

The relationship between the box dimension and areographical fragmentation can be further clarified by considering the relationship between the fractal dimension of the perimeter of a geographical range and the box dimension. Suppose we want to measure the absolute size of a geographical range. Since the boundary of the geographical range is fractal, the size we calculate will depend on the size of the ruler we use. However, we may obtain an approximation by choosing a window of size h, counting the minimum number of these windows (boxes) that are required to cover the geographical range, and multiplying the count by h^2. Thus, the estimate of area, $A(h)$, will be

$$A(h) = N(h)\, h^2$$
$$= a\, h^{2-D_B} \tag{3.7}$$

where a and D_B are obtained from a regression as in equation 3.6. Now the perimeter is also a fractal, so we can use equation 3.5 to express this relationship as

$$P(h) = p \left(\frac{1}{h}\right)^{D_P}$$

where D_P is the fractal dimension of the perimeter. If this equation is solved for h and the result substituted into the previous equation, we have

$$\frac{A(h)}{a} = \left[\frac{P(h)}{p}\right]^{\frac{2 - D_B}{-D_P}}$$

Solving for D_B gives

$$D_B = 2 + D_P \left(\frac{\log A(h) - \log a}{\log P(h) - \log p}\right) \tag{3.8}$$

But $\log A(h) < \log a$, so the box dimension decreases for increasing convolution or jaggedness of the perimeter. Notice also that the relationship is independent of scale, since the ratio in the second term does not explicitly include the window size: an arbitrary pair of areas and perimeters *measured at the same scale* can be used.

Equation 3.8 primarily expresses the relationship between the area and the perimeter of the geographical range of a single species. Area−perimeter relationships can provide additional information about the variability of the geographical range areas and perimeters among different species. Lovejoy (1982) discovered that for shapes made by cloud shadows, perimeters were related to shadow areas of different clouds as

$$P \propto A^{\frac{D_P}{2}}$$

where P is the perimeter of the cloud shadow and A is the area. If the shadow were a Euclidean shape such as a square or circle, $D_P = 1$, and the perimeter would increase with the square root of area. That is, perimeter rises proportionately more slowly than area. A circle with twice the area of another would have only 1.41 times the perimeter. The fractal dimension D_P is a measure of how fast perimeter rises with increasing area. For irregular shapes, with many peninsulas and extensions, $1 < D_P < 2$. The higher the fractal dimension, the more rapidly perimeter increases with area.

The application of this to geographical ranges is as follows. The edge of a species' geographical range marks the limits to its tolerance of extreme ecological conditions (e.g. temperature, Root, 1988a). Species with large geographical ranges will have a proportionately lower amount of perimeter than species with small geographical ranges. Thus, a species with a large geographical range will have relatively fewer local populations found at the extreme of ecological conditions for the species. The higher the fractal dimension associated with area—perimeter relationships for a group of species, the faster the perimeter increases with increasing area, and consequently, the proportional difference in perimeters between species with large geographical ranges and those with small ones will be smaller. Hence, the beneficial effects of having a large geographical range, i.e. having relatively fewer local populations in contact with extreme conditions for the species, will be lessened in a species that has a high value for D_p. The consequences of this for the persistence of a species across its geographical range could be profound. Populations that are in contact with environmental extremes may have a higher chance of becoming extinct than those in landscapes with more moderate conditions.

3.2.3 Example analyses using BBS data

The BBS data were used to draw geographical range maps of a number of species of North American terrestrial birds. These range maps represent areas of occupancy of the species because data on population abundances were used to draw boundaries where abundances dropped to zero. I will discuss the details of the techniques used to obtain these maps in the next chapter. Here, I use them to illustrate how to calculate box dimensions. I also describe the geographical range area—perimeter relationships obtained from these maps.

The maps in Fig. 3.6 show the geographical range of Baird's Sparrow (*Ammodramus bairdii*). This sparrow breeds in the northern Great Plains region of North America, and migrates south to south-western Texas and southeastern Arizona during the winter. There were some sightings of the species in eastern New Mexico during the breeding season for at least one or two years. Hence, a small patch is shown on the maps far to the south of the rest of the breeding range. I included the patch in the maps even though Baird's Sparrow does not normally occur there to reflect that there were some birds found outside their normal geographical area. I did calculations with the map both including and excluding the small patch, and there was little difference in the results.

To illustrate how to go about measuring the box dimension, I have superimposed grids of different sizes on top of the range. The window

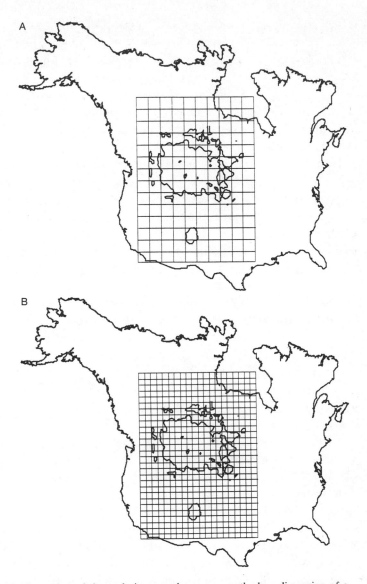

Fig. 3.6 Illustration of the technique used to measure the box dimension of a geographical range. The two grid sizes are used to cover the geographical range, and the minimum number of boxes in each grid that cover the range are counted (the reader is invited to count boxes to develop an appreciation for modern computers, which can count much quicker). Note that the grid may need to be adjusted to achieve the minimum number, the ones illustrated here are assumed to be minimal box grids. The geographical range illustrated here is that of Baird's Sparrow.

size of the grid in (A) is twice as large as the grid in (B). Using a computer one would adjust these grids so that the smallest number of boxes would be covering the geographical range. Here, for illustrative purposes, we will assume that the grids shown are the minimum covering grids. In (A), I counted 59 (54 without the southern patch) boxes covering the geographical range, and in (B) I counted 172 (163 without the southern patch). One would expect that since 4 smaller squares fit into each of the larger squares, then there should be 4 times as many of the smaller squares covering the range than of the larger squares. However, 172 divided by 59 is around 3, so there are only 3 smaller squares covering the range for every large square. The smaller squares have a greater chance of not containing part of the geographical range than the larger squares near the border. The rate at which smaller squares miss the border as the size of the square decreases is a measure of the jaggedness of the border. This rate is estimated by the slope of the log−log regression in equation 3.6.

The process of counting how many boxes of different sizes it takes to cover the geographical range of Baird's Sparrow was left to a computer algorithm. The results of these calculations are shown in Fig. 3.7. I have also included the results obtained from identical calculations for two congeneric species: the widespread and abundant Grasshopper Sparrow (*Ammodramus savannarum*) which breeds nearly from coast to coast in appropriate habitat and the very localized Henslow's Sparrow (*Ammodramus henslowii*) which breeds throughout most of north-eastern North America but is often very sporadic in its occurrence. The geographical ranges of these species are shown in Fig. 3.8. The slope of each of the regression lines in Fig. 3.7 are estimates of the box dimension of the corresponding geographical ranges.

Notice that there is very little scatter about the regression lines (R^2 values for each regression are > 0.95). Since spatial autocorrelation exists in each data set it is not possible to use conventional statistical tests to compare slopes. But examining the ranges of the species and comparing them to the corresponding box dimensions illustrates the potential usefulness of this measure of fragmentation. The geographical ranges of Baird's and Henslow's Sparrows are approximately the same size (the intercepts of their respective log−log regressions in Fig. 3.7 are nearly equal), but Henslow's Sparrow seems to have a more fragmented and convoluted range boundary than Baird's Sparrow. The box dimension for Henslow's Sparrow is 1.57, that for Baird's is 1.62, indicating that the geographical range of Henslow's Sparrow is less space-filling than that of Baird's. Baird's Sparrow just happens to

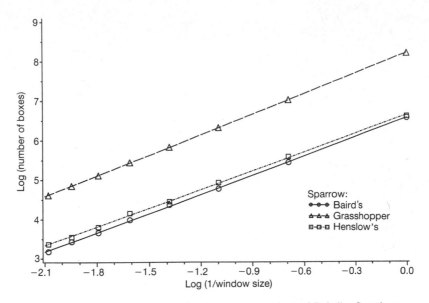

Fig. 3.7 Regressions used for estimating the box dimensions of Baird's, Grasshopper, and Henslow's Sparrows' geographical ranges. Note that the slope for Henslow's Sparrow is the shallowest and that for the Grasshopper Sparrow is the steepest. The Grasshopper Sparrow has the largest geographical range because its intercept is much higher than the other species. Logarithms are base e.

be more abundant on the average than Henslow's Sparrow. The Grasshopper Sparrow's box dimension is 1.72, indicating that its geographical range is more space-filling than either of the other two sparrows. It also has a clearly larger geographical range than the others, with breeding populations across most of the continent except for the northwest. Thus, since there are few large 'holes' in its geographical range, there is a greater amount of area for a given length of border of the geographical range for the Grasshopper Sparrow. These comparisons between the three sparrow species are certainly not definitive, but they suggest that the box dimension can be a useful statistic for describing the complexity of the shape of a species' geographical range.

The analyses in the previous paragraphs of this section focused on the box dimension of a species' geographical range. Another way of looking at this problem is to examine area−perimeter relationships (see equations 3.7 and 3.8). I have plotted the logarithm (base 10) of geographical range area against the logarithm of geographical range perimeter measured using a window size of 40 km (about the length of a BBS route) for 85 species of passerine birds in the taxa Tyrannidae

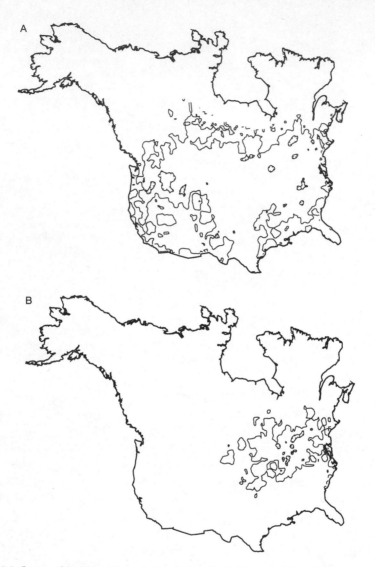

Fig. 3.8 Geographical range boundaries of Henslow's Sparrow (A) and Grasshopper Sparrow (B) obtained from the BBS data set. Boundaries represent estimates of locations where abundance drops to zero.

(flycatchers), Vireonidae (vireos), Icterinae (blackbirds, orioles and allies) and Parulinae (wood warblers) in Fig. 3.9. Some of these species migrate to the neotropics, others remain in North America during the winter. Species of neotropical migrants have a significantly different perimeter–area relationship than temperate migrants and residents (hereafter called residents). The fractal dimension of the perimeter–

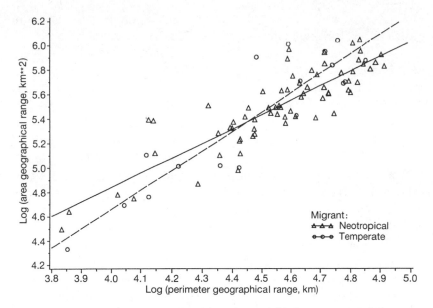

Fig. 3.9 Relationship between the logarithm of geographical range area and the logarithm of geographical range perimeter for 85 species of North American passerine birds. The two regression lines were fit using ordinary least squares. For neotropical migrants the slope is 1.18; for residents, it is 1.58. These are significantly different ($t =$ 2.32, d.f. $= 82$, $P = 0.02$).

area relationship for neotropical migrants is 1.69, that for residents is 1.26. Thus, the relative advantage of having a large geographical range is less for neotropical migrants than for residents.

One can also think of the area–perimeter relationship in a dynamical sense. For a species that follows a particular area–perimeter law, as its geographical range increases in size, it will cover relatively more area and have a relatively less convoluted boundary. Hence, it will begin to have relatively fewer populations that experience harsh or extreme conditions for the species. Just the opposite will occur as the geographical range of a species decreases. The smaller its geographical range becomes, the higher proportion of populations come into contact with environmental extremes. This process would happen faster for a neotropical migrant than for a resident. Clearly, this has implications for the management of neotropical migrant songbirds.

3.3 Summary
Areographical properties of geographical ranges deal with the size, shape and orientation of the range, and the nature of its boundary. A

number of measures of size, shape and orientation are available from basic data on the number of sites at which a species is known to occur. Using such data, the centroid of the geographical range and variances and covariances with respect to geographical position can be calculated. If data on the abundance of the species is also available at these sites, then these positional statistics can be weighted by abundance, so that geographical positions with higher abundances add more information to the calculation. From these basic positional statistics, measures of geographical range shape and position can be calculated using major and minor axes of an ellipse centred on the geographical range centroid. It is also possible to calculate a relative measure of the area of occupancy of a geographical range by calculating the area of an ellipse with a standard relative size of the major and minor axes. Data from the BBS indicate that these measures suggest that some interesting patterns emerge from their analysis.

Geographical range boundaries are discontinuous at many scales, and thus the outline made by a geographical range can be considered to have a fractal shape. This shape can be described by calculating a fractal dimension. The box dimension of the range boundary describes the space-filling properties of the shape. The smaller the box dimension, the poorer the space-filling properties of the geographical range, and consequently, the more convoluted the boundary becomes. Comparisons of the degree of geographical range fragmentation among groups of species can be made by considering the relationship between range perimeter and range area, as measured by a common scale. Analyses of BBS data indicate that neotropical migratory species tend to have a different perimeter−area relationship than their relatives that do not migrate to the neotropics.

The techniques described here can be used to analyse questions regarding how the size, shape and orientation of geographical ranges might affect the processes that determine biological diversity. Brown and Maurer (1989) argued that diversification of biotas is greatly influenced by the physical nature of the continent. This influence can be seen by considering the statistical distribution of species along axes representing different properties of their geographical ranges. Geographical range size is also likely to be associated with extinction likelihood, particularly for species that have relatively small geographical ranges.

CHAPTER 4

Analysis of geographical variation in abundance

Population densities of species are not constant across their geographical ranges (Brown, 1984; Maurer & Villard, 1993). This was clear to naturalists in Darwin's day, and to a certain extent entered into his thinking on natural selection. It influenced even more his ideas on how the evolutionary process worked. But since the synthesis of genetics and Darwin's evolutionary theory, studies of patterns in the distribution of abundance in geographical space have played a relatively minor part in our thinking about how evolution occurs and how species react to their environments. As I discussed in Chapter 1, interactions among species' populations and their resources across geographical space may have profound consequences for our ability to conserve them in the face of drastic changes in global ecosystems. Techniques for analysis of large data sets on population abundances are available from literature on geostatistical methods, but these methods are not yet in general use among ecologists and biogeographers.

A number of statistical methods exist that can be used to describe spatial patterns in ecological systems (see, for example, Ord, 1979; Legendre & Fortin, 1989). Not all of these techniques are appropriate for the study of geographical populations. Studying patterns of spatial autocorrelation may be useful in geographical population analysis (see Rossi *et al.*, 1992), however, the study of spatial point patterns may be much less informative. Samples of geographical populations tend to be made from irregularly spaced points on a geographical grid, and such a sampling protocol precludes the use of certain kinds of statistical techniques. Many statistical techniques developed for use in ecology are designed to analyse samples obtained at relatively localized scales (e.g. within a community). Assumptions underlying many spatial statistical tests may be violated when trying to generalize them to geographical scales.

The goal of this chapter is to introduce some techniques for analysing the spatial pattern in abundances of species and for developing methods of describing demographical fragmentation of a species. Such methods are necessary in order to develop tests of hypotheses about how spatial patterns of abundance influence the likelihood of persistence of a species. Extinction is the end result of a consistent statistical pattern of population decline and geographical range shrinkage. Ultimately, what is needed are techniques that can be used to identify spatial patterns in current geographical populations of species that might be used as indicators of widespread population decline in the species. This presupposes an ability to describe spatial patterns in populations and to relate those patterns to population changes over wide geographical areas. In this chapter, I focus on the first of these problems: the description of geographical patterns of population abundance; in the next chapter I will take up the problem of describing spatial patterns of population change. I first discuss methods of using randomly sampled locations across a geographical area to generate contour maps that describe the variation in abundance of a species across space. I then consider methods for summarizing the information on such contour maps that is relevant to demographical fragmentation. It is this kind of fragmentation that may be most closely related to population decline.

4.1 Estimating spatial trends in population abundance

L.R. Taylor was one of the first ecologists to examine the problem of estimating spatial patterns of abundance in populations (see his review in Taylor, 1986). His work spanned nearly three decades and incorporated not only statistical models of abundance, but also theoretical models that attempted to relate the behavioural characteristics of individual organisms to the spatial and temporal patterns exhibited by their populations across their geographical ranges. Taylor's pioneering contributions laid the foundation for future progress in this area.

The model of populations of animals and plants being discontinuous at multiple scales makes the estimation of spatial trends in population abundance a particularly difficult problem to deal with. Any representation of the distribution of abundance of a species across its geographical range as a smoothly varying surface will have limitations of the scale of applicability. Before discussing methods for examining spatial trends in abundance, it is important to examine some of these limitations.

If censuses are done across a continent at random locations, it is inevitable that some of these locations will contain large portions of some metapopulations while others will contain small portions or even no parts of metapopulations. Hence, at the geographical scale, different census locations will contain different amounts of the species' specific habitat elements required to maintain viable metapopulations. If a statistical technique attempts to interpolate an abundance measurement between two randomly sampled locations, then it is important to understand how such interpolated measurements are to interpreted.

The first thing to realize about interpolated measurements is that the same interpolated trend may correspond to a number of different processes that might be responsible for the estimated changes in abundance. The scale of the original data will limit the ability of the researcher to distinguish between mechanisms. For example, suppose an interpolation among sampled locations reveals a trend of decreasing population abundance across some geographical region. There may indeed be a tendency for populations to decrease across the region, but it would be impossible to say whether that decline represented a decrease in the amount of adequate habitats available or a decline in population density within habitats but no general reduction in the amount of habitat (i.e. a reduction in habitat quality rather than quantity). Therefore, interpolated maps will not be useful in testing hypotheses about specific mechanisms responsible for spatial variation in abundance. They will, however, provide a basis for comparison among species of spatial patterns in abundance, provided the same criteria are used to construct maps for each species to be compared. These criteria include both the sampling protocol used to obtain the data and statistical decisions made in constructing the maps.

Contour maps obtained from interpolations always have a characteristic scale. By scale I am referring to the size of the smallest geographical unit recognized by the interpolation procedure. This means that points in geographical space that are closer to one another than the scale of the map will be assigned the same abundance by the interpolation procedure, regardless of the number of individuals actually existing at each point. Thus, the interpolation procedure will probably over-estimate abundance in some geographical regions while under-estimating abundance in other regions. A good interpolation procedure will not bias interpolated measurements one way or the other, and will reflect broad geographical patterns in changes in abundance rather than abundances at individual locations. The value of the interpolated map is in its ability to provide a common basis for broad scale comparisons among species of spatial changes in abundances rather than specific statements about what to expect at localized points in geographical space.

4.1.1 Spatial trend surface analysis

There are a number of methods that have been developed in the geostatistical literature for constructing maps describing quantitative trends in continuous variables that change over space. One of the original methods used for constructing such maps was models based on polynomials in two variables (i.e. two spatial dimensions). All methods for developing spatial trend surfaces are computationally intensive and therefore require access to relatively fast, powerful computers, especially when a large amount of data is to be analysed. With the advent of geographical information systems (GIS) technology, the potential to use these methods to analyse data from geographically extensive populations censuses such as the BBS now exists. The following discussion highlights the detailed treatments of these methods found in Ripley (1981), Burrough (1986), Davis (1986), Sampson (1988), Haining (1990) and Cressie (1991).

Perhaps one of the simplest procedures for obtaining an estimate of spatial trend is the use of what are called 'moving averages'. Assume that a set of N sites have been censused, and an estimate of population abundance is obtained at each of these sites. Let these data be denoted as $x_1, x_2, \ldots, x_N$. A moving average for a point not included in this sample can be obtained by choosing a window size, and using all points in that window to calculate an average. Thus

$$x_j = \frac{\sum_{i=1}^{n(h)} x_i}{n(h)}$$

where $n(h)$ is the number of points in a window of size h around the point to be estimated. The window is then 'moved' systematically across the surface to be interpolated, and sampled points are included in the moving average estimate for points to be interpolated until a regularly spaced grid of values are obtained that can be used to represent the surface.

The selection of a particular window size is somewhat arbitrary, so it is sometimes desirable to pick some type of objective criteria for determining window size used in a moving average. Points to be included in the moving average are weighted using some weighting rule, so that the interpolated estimate can be given by

$$x_j = \frac{\sum_{i=1}^{n(h)} w_i x_i}{\sum_{i=1}^{n(h)} w_i}$$

where $n(h)$ is now the number of points for which $w_i > 0$ (negative weights are not used). One of the most common weighting functions is distance, where $w_i = d^{-t}$, $t > 0$, and d is the distance between a sampled point and the point to be interpolated. The effective window size decreases as larger values of t are used in the moving average.

Weighting procedures use objective, yet arbitrary, criteria to include points in the interpolation process to estimate a spatial pattern from sampled locations. There is a procedure based on the GRVM that circumvents this problem by using information on the patterns of spatial correlation among sampled locations to estimate interpolated points. The procedure is called 'kriging' (e.g. Ripley, 1981; Burrough, 1986; Cressie, 1991) and incorporates information from the semi-variogram discussed in Chapter 2 to obtain the weights. Kriging is used to estimate weights as follows. Let $\gamma(x_i, x_j)$ be the semivariance between points x_i and x_j. Then if x_k is the point to be interpolated, the weights assigned to each observed point used to interpolate x_k are obtained by solving the following set of equations

$$\sum_{j=1}^{N} w_{jk}\, \gamma(x_i, x_j) + C = \gamma(x_i, x_k) \tag{4.1}$$

for $i = 1, 2, \ldots, N$. The constant C is required to ensure that the w_{jk}'s all sum to unity. Assuming that the semivariance function is known, then *for each point to be interpolated*, one must solve N equations with N unknowns (i.e., the w_{jk}'s). Usually, if there is a well defined range to the semivariogram, then it is not necessary to include data from points beyond the range of the semivariogram, which are considered spatially independent. Nevertheless, kriging is computationally very intensive and time-consuming, requiring fast computers with plenty of memory. Once the weights have been estimated from equation 4.1, then the estimated value of the point is

$$x_k = \sum_{i=1}^{N} w_{ik}\, x_i$$

One major problem with kriging is that the semivariances cannot be estimated from equation 4.1, but are assumed to be known for an arbitrary distance between points. This difficulty is circumvented by fitting a parametric function to the estimated semivariogram given by equation 2.3. A number of different models have been used for this purpose. Here I review several commonly used models.

The first semivariance model used is a linear model. In the linear model, the semivariance is simply a linearly increasing function of lag distance. If a nugget effect exists, then the semivariance can be written as

$$\gamma(h) = c_0 + a\, h \tag{4.2}$$

where c_0 is the nugget variance. This version of the linear semivariogram does not implicitly include a sill. This may be the case, for example, if the extent of the data set does not cover the entire support of the geographical range of a species. This might be the case when a census programme ends at a political boundary, but a species' geographical range does not. The linear semivariogram model can be modified to include a sill.

The spherical semivariogram model is

$$\gamma(h) = c_0 + c_1 \left[\frac{3h}{2a} - 0.5 \left(\frac{h}{a} \right)^3 \right] \tag{4.3a}$$

when $h < a$ and

$$\gamma(h) = c_0 + c_1 \tag{4.3b}$$

when $h > a$. The parameter a is the range and $c_0 + c_1$ is the sill. This model is often used in geostatistics. A model that gives similar results is the exponential model

$$\gamma(h) = c_0 + c_1 \left[1 - \exp \left(\frac{-h}{a} \right) \right] \tag{4.4}$$

These models can be fit to the data using nonlinear or weighted least squares models. If a particular distribution is appropriate for the data, then these models can be fit using maximum likelihood estimation (Burgess & Webster, 1980a,b; Haining, 1990). A number of other semivariogram models have been used, and the relative abilities of the models to fit data can be assessed using conventional goodness of fit statistics, such as the R^2 value obtained from the least squares analysis. There are often a number of practical concerns in kriging, such as the number of points used to obtain the weights in equation 4.1. These are discussed in manuals for the use of computer programs that calculate kriging estimates of spatial data, such as SURFACE III (Sampson, 1988) or GS+ (GS+, 1991).

Another set of interpolation techniques uses a series of planes in the neighbourhood of a grid point to interpolate values at that point from irregularly sampled data. For example, Root (1988b) used a least squares regression plane to interpolate the value at grid cells in the neighbourhood of at least four and not more than 24 sampled points from the Christmas Bird Counts. These methods are often highly flexible and can provide useful spatial trend surfaces. Trend surface analysis based on these methods can be done in the SURFACE III program (Sampson, 1988).

4.1.2 Example analyses: geographical distribution of Eastern Kingbird abundances

Using the BBS data set, I extracted all censuses on which Eastern Kingbird (*Tyrannus tyrannus*) was observed. This species is a widespread tyrant flycatcher that nests in trees in open habitats and is very aggressive when defending its nest. For each of 3017 BBS censuses, the abundance of Eastern Kingbirds was either zero if the species had never been seen on the census, or if it had, its abundance was calculated as the average over all years that the species was observed on the census route. These data formed the basic data set used to obtain a spatial trend surface for the species.

Two different methods were used to obtain the spatial trend surface. Firstly, I used a weighted moving average algorithm that selected the eight censuses nearest to each point on a regularly spaced grid with a grain size of approximately 40 km (i.e. the distance from each node on the grid was about 40 km). The values of each census point were weighted by the inverse of the squared distance between the grid node and each of the eight nearest censuses. Thus, the effective window size for each grid point varied depending on how close the nearest censuses were.

The second method used to obtain an estimate of the trend surface was kriging, outlined in the last section. Firstly, the sample semivariogram was calculated using equation 2.3 and the 3017 census points (remember, some censuses had zero counts). The sample semivariogram is shown in Fig. 4.1A. Note that it is not a 'well behaved' semivariogram, that is, it does not increase then level off at a well defined sill. Instead, it levels off briefly, then again increases to reach a peak at about half the maximum lag, then declines again. This is because geographical ranges have the well defined deterministic property of abundances declining away from the geographical range centre (Brown, 1984; Maurer & Brown, 1989; Hengeveld, 1990). Returning to the definition of the semivariance in equation 2.2, this means that the spatial process that determines the geographical distribution of abundances is not stationary, so that the drift is nonzero. If we assume that the deterministic process of declining abundances away from the range centre manifests itself at relatively large geographical scales, then the variation in abundances may be stationary at small geographical scales. Inspection of the semivariogram bears this out. If semivariances at lag distances greater than one third of the maximum lag are deleted from the analysis, a 'better behaved' semivariogram is obtained (Fig. 4.1B). We have found in doing many such calculations (see Maurer & Heywood,

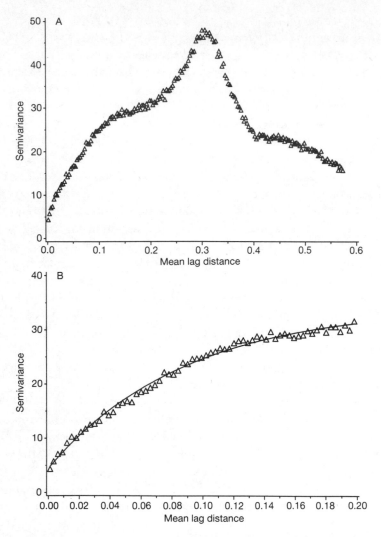

Fig. 4.1 Semivariogram for the Eastern Kingbird obtained from BBS data. (A) shows the entire semivariogram. Notice that the semivariogram reaches a peak at about half the maximum lag distance, then declines. (B) shows only the first 25% of the semivariogram. Over these relatively short spatial scales, it is clear that the semivariogram looks like that of a stationary process. The best fit exponential model is given by the solid curve. It was estimated using nonlinear least squares.

1993) that the semivariograms of most species tend to look like that of a stationary spatial process at small lag distances.

To complete the calculation of the spatial trend surface, the exponential semivariance model was fit to the sample semivariogram for the Eastern Kingbird data (Fig. 4.1B). The sample semivariogram and the fitted exponential model were calculated using the program GS+ (GS+, 1993). The trend surfaces for both methods were calculated using the SURFACE III program (Sampson, 1988) on an IBM RS6000 UNIX workstation. This computer has plenty of memory and is very fast, so calculations took only minutes instead of hours (as they would on our fastest PC, a 33 mhz 486 machine).

The maps generated by the weighted moving average technique and kriging were strikingly similar in their overall appearance (Fig. 4.2). This lends some support to the objectivity of the spatial pattern of abundance for Eastern Kingbirds documented by the BBS, since the pattern does not seem to be an artefact of the technique of estimating the spatial trend. Both maps indicated that there was a region of maximum abundance extending along a north–south axis through the Great Plains, with the highest abundances occurring in North Dakota and southern Saskatchewan. A second region of high abundance, but not as high as the first region, occurs along the Great Lakes and in New England. A third region of high abundance is found in the southeastern United States. The major difference between the two maps is that the one obtained from kriging tends to have more jagged borders than the one obtained by moving averages. If the weighting function were to be changed for the moving average map, then borders could be made more jagged.

Since the maps agree in the major features of the distribution of abundance for Eastern Kingbird, it is useful to compare the strengths and weaknesses of the two approaches. When considering computer time and access, the advantage clearly belongs to the moving averages technique, since this technique is not nearly as computer-intensive as kriging. However, it has the disadvantage of requiring an arbitrary choice of the weighting function used. There are no theoretical grounds that would provide an a priori value for the exponent for the distance weighting function. Kriging uses the semivariance to determine those weights, so that patterns of spatial autocorrelation are used explicitly. This has the advantage of incorporating information on spatial trends into the kriging estimates. However, since kriging requires the assumption of stationarity, it is necessary to examine the semivariogram carefully in order to identify lag distances over which the spatial trends in population abundance are stationary. There are few objective criteria that can be used to make this determination. A way around such problems is to use a technique called universal kriging. This method does not assume that the spatial process being modelled is stationary (see Cressie, 1991 for further detail).

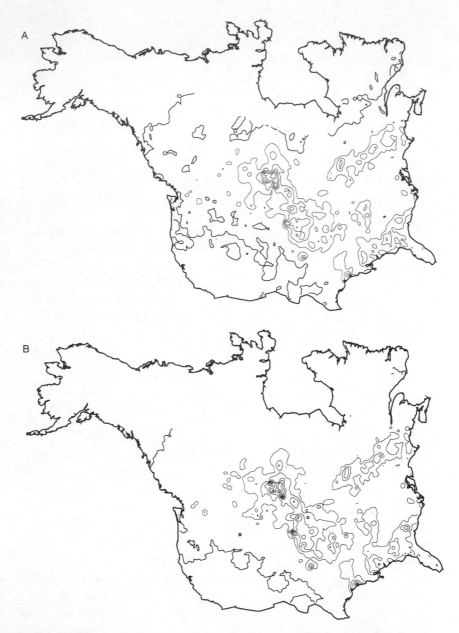

Fig. 4.2 Maps showing the estimated geographical distribution of abundances for Eastern Kingbirds generated using two different techniques. (A) was generated using a weighted moving average algorithm, where weights were proportional to the inverse of the squared distance from a sample point to the point to be interpolated. (B) was generated using kriging. The semivariance function used is shown in Fig. 4.1B. Bold contours indicate lines of zero abundance. Lighter contours connect points with equal abundances. Contours are spaced at intervals equal to 15% of the maximum abundance.

4.1.3 Calibrating population abundance maps
with data on population density

An important consideration regarding abundance maps, such as those described in the previous section, is how well the results of extensive surveys, such as the BBS, reflect local population densities measured on intensive, but spatially restricted plots. There are a variety of reasons to suspect that systematic biases could exist in census techniques such as the BBS that would cause BBS estimates of abundances to fail to represent population densities of individual species. I will consider a few of these briefly before describing a procedure that can be used to calibrate abundance maps using intensively obtained estimates of population density. Such estimates are often more accepted by biologists than the results of extensive surveys.

When sampling populations over large regions, there is often a trade-off that must be made between the time spent at each location sampled and the time allocated to sampling more units. For each additional amount of time spent at one location, there are fewer additional locations that can be censused. A potential problem with this trade-off is that extensive sampling techniques may miss detecting species that for some reason are difficult to detect without spending a sufficiently long period of time in the census area. A second problem with extensive sampling is that standardization of census techniques used to detect species at each point may bias the ability to count certain species. For example, in the BBS, routes are censused from sunrise to late morning. Some species, such as nocturnal raptors, may not be represented in their actual abundances because the time of day the census is run biases against detecting them. Another problem with extensive surveys is that in order to be practical, they must be located along routes that make transportation from one site to another practical. For the BBS, this means that counts must be conducted along roadsides. Abundances of species in habitats that are rare along roadsides, or that are affected by the presence of the road, such as deep forest interiors, may be consistently under-estimated. A number of specific concerns are discussed regarding bird censuses by papers in Ralph and Scott (1981).

One way to get some idea of the kinds of sampling biases that might be incorporated into extensive censuses is to make some comparison between abundances of species obtained from intensive surveys and those made from extensive surveys. This can be done using the trend surface methods described in the previous sections. Let y_i be the density obtained for a species on the ith intensive census, and let x_i be the estimated abundance of that species obtained from a trend surface map. It is unlikely that an intensive survey will be contained in the exact area in which an extensive survey is conducted, but if this is the case, for all intensively censused locations, then the data obtained from the extensive survey may be used instead of the value estimated by the trend surface. The relationship between the extensive and intensive surveys can be estimated using a linear or nonlinear regression model or a nonlinear curve fitting routine to explore possible quantitative relationships between the x_is and the y_is. Notice that the regression procedure is used only to estimate the relationship between densities obtained from the intensive surveys and counts from the extensive surveys. Since values of each of these variables will be autocorrelated in space to a certain degree, conventional statistical hypothesis tests will not be valid. The goal of the exercise is to estimate the relationship rather than to test hypotheses, and since the regression estimates will be unbiased even in the presence of autocorrelation, the regression procedure will provide a valid way to relate counts.

Once a quantitative relationship has been estimated between the intensive and extensive counts, then that relationship can be used to 'calibrate' the surface obtained from extensive counts. That is, if the regression or curve fitting procedure indicates that there is a sufficiently strong quantitative relationship between the intensive and extensive censuses, the surface can be expressed in terms of densities rather than abundances. An advantage of this procedure is that one can also examine the residuals from the regression spatially to identify specific areas in the geographical range where abundance estimates may poorly reflect densities.

I now illustrate the calibration technique using data from BBS counts to obtain trend surfaces and compare those surfaces with density estimates obtained from the Breeding Bird Census (BBC) programme. The BBC programme has been coordinated by the Audubon Society for over 50 years. Individual censuses are obtained by intensive study of relatively small (up to 10 ha) plots. Observers use the 'spot mapping' technique first described by Kendeigh (1944). Spot mapping consists of repeated censuses of an area to identify the locations of males singing on their territories. Although there are some limitations to this technique, it is often considered the most reliable estimate of breeding birds in a location, and often serves as the basis for comparison with other bird census methods (see papers in Ralph & Scott, 1981). In Fig. 4.3, the abundance trend surface for Eastern Meadowlark (*Sturnela magna*) obtained from the BBS is given. Note that there is a high concentration of abundance in southeastern North America for this species.

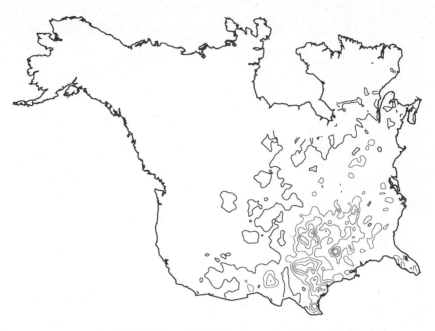

Fig. 4.3 Geographical distribution of BBS abundances for Eastern Meadowlarks obtained from kriging. Note the concentration of abundances in the southeastern portion of North America.

Densities of Eastern Meadowlarks were obtained from 99 BBCs conducted after 1965. The Cornell Laboratory of Ornithology serves as the repository of these censuses, and maintains them in a computer data base. When abundances estimated from the BBS trend surface were plotted against the densities obtained from the BBC (Fig. 4.4A), a weak but significant correlation was obtained ($r = 0.32$, d.f. $= 98$, $P < 0.01$). There was a great deal of variability evident in the relationship between the two measures of abundance.

The data for the Eastern Meadowlark in Fig. 4.4A indicate that there may be a nonlinear relationship between BBS counts and BBC densities. Taking logarithms for both measures of abundance did not improve the linear relationship, in fact it decreased the correlation coefficient by about half. I explored possible nonlinear relationships using a curve fitting routine. Firstly, because the data are likely to be subjected to considerable measurement error, I used a smoothing routine to obtain a smoother relationship between the abundance measures. The general idea is to smooth out some of the potential noise in the data while retaining its basic patterns of variation. Several algorithms are available for this purpose. I tried several smoothing routines including fast fourier transform filtering, polynomial interp-

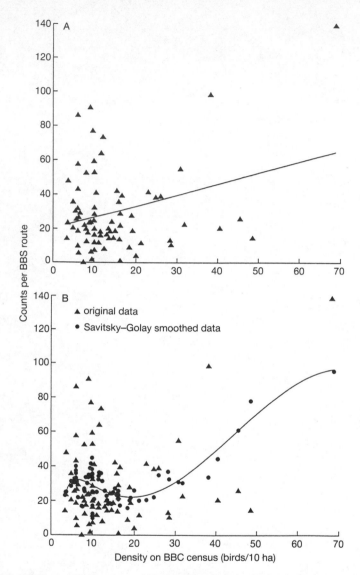

Fig. 4.4 (A) Plot of BBS estimates of abundance against densities obtained from BBC locations. The line was estimated by ordinary least-squares. (B) Nonlinear analysis of BBS and BBC data. The curve given is a nonlinear model fitted to smoothed data (circles). The original data are given for reference. (There are fewer data points in this plot because BBC data from the same site were averaged over years).

olation, locally weighted regression (lowess), and least squares quartic polynomial fitting with a moving window (Savitsky—Golay). Several of these routines gave very similar patterns of smoothing, and I used one (Savitsky—Golay) because of its computational efficiency. The smoothed data are given with the original data in Fig. 4.4B. A large number of nonlinear models were evaluated that gave essentially identical patterns to the curve shown in Fig. 4.4. These calculations were done using the TableCurve program (Jandel, 1992).

Because of the large degree of variability of the original data, it is perhaps a little tentative to read much into the analysis given in the lower panel of Fig. 4.4. However, I think that there are some qualitative statements that can be made about the relationship between BBS and BBC abundances of the Eastern Meadowlark. There appears to be a range of densities (about 0−5 birds per 10 ha) when meadowlarks are rare over which the BBS abundance estimate rises rapidly with BBC density. Then, there is a rather large window of densities (5−40 birds per 10 ha) over which BBS abundances are relatively insensitive to underlying densities. Finally, when meadowlarks are very abundant, there is again a rise in the number of birds seen on BBS routes as underlying density increases. This suggests that the BBS is relatively insensitive to variations in density when meadowlarks are moderately abundant. Another point worth mentioning is the relationship between the original data and the smoothed data. If the deviations between these two sets of data actually represent sampling error, then it is obvious that at least for the BBS, there is a considerable amount of variability in estimates obtained through the current BBS sampling protocols. Nonlinear smoothing of the BBS data using BBC estimates as a calibration data set may provide a means for dealing with some of this sampling variability.

This exercise was repeated for several other species, and the results were generally similar. It seems likely from these preliminary efforts to relate BBS and BBC measures of abundance that both census methods have some limitations. BBCs are intensive censuses of small plots. It is very possible that in some situations BBC densities will not be representative of the abundance of a species in a landscape, particularly if the census plot happens to be located in a unique patch of habitat that does not represent the 'average' conditions of the habitat in a landscape. On a geographical scale, BBCs are located relatively haphazardly, the primary determinant of the location being the desire of an ornithologist to spend the time to collect the data. BBS routes, on the other hand, are located extensively across the continent, with attempts made to make coverage of the continent as complete as possible. But the method of censusing birds used in the BBS does not account very well for variation in local abundances within landscapes. Generally, the 40 km long BBS route will often cross several different landscapes. Even if abundance is not significantly biased by the location of a route along roads, BBS censuses will tend to smooth out variations in local abundances on the scale of the 10 ha census plot. BBS census data may be adequate to measure the spatial trends in average abundance across geographical space, but they may not provide data on spatial trends in *local densities*.

It may be possible that BBS and BBC abundances are both accurate, yet at the same time not highly correlated. To see this, picture a species that is limited to a certain kind of habitat that occurs in patches in a landscape. In the centre of the species' geographical range, the patches of appropriate habitat might occur regularly within regional landscapes, so that its abundance on a BBS route might be relatively high, even if its density within patches does not vary much. If patches of the species' favoured habitat are less abundant within landscapes at the edge of the species' geographical range, its BBS abundances may be low, even if it is found at roughly the same densities within patches as it would be in the centre of its range. In such situations, BBS- and BBC-type censuses would not necessarily be correlated. Rather, both would provide different kinds of information about the species' abundance. How common this situation is will require a much larger sample of species than the few analyses done here.

4.1.4 Relating abundances to environmental variables
One of the reasons that obtaining trend surfaces is likely to be extremely useful in conservation efforts is that they can be used to relate population abundances to environmental variation across continents. There have been major logistical difficulties with attempting such correlations in the past. One of the few studies to attempt continental-level analyses of habitat and bird populations was done by Emlen *et al.* (1986). Their study was only able to cover a small part of the North American continent, along the Mississippi River.

With the advent of remotely sensed environmental data, collecting data on spatial variation in environmental data has been greatly enhanced. Large areas of continents can be sampled at relatively small scales. The major problem with such data is resolution. Most of these data provide only coarse descriptions of wildlife habitat. At present, plant species' composition cannot be assessed, although different life forms (e.g. grasses, shrubs, trees) can sometimes be identified by properties of reflected light (e.g. more light will be reflected by grasslands than by a dense canopy of trees) and seasonal patterns of productivity.

In order to assess the kinds of statistical problems that might be encountered in correlating continental patterns of bird distributions and vegetation, geographical data on variation in remotely sensed vegetation patterns were obtained from the Global Ecosystems Database, Version 1.0 (Kineman & Ohrenschall, 1992). A 1° grid of June average values of the Normalized Difference Vegetation Index (NDVI) was constructed from this data base for North America. Values for this index were plotted against abundances for two bird species. Baird's Sparrow (*Ammodrammus bairdii*) is a relatively restricted species with a small geographical range centred in the middle of North America. It has a single region of high abundance that extends longitudinally across the centre of its range. American Redstart (*Setophaga ruticilla*) is a widespread and abundant species that has an extensive geographical range in eastern and northern North America. There are two concentrations of abundance for this species, one centered in eastern Canada, and one along the Pacific coast of Canada and southern Alaska (see Fig. 5.13).

Plots of abundance against June NDVI indicated that regions of peak abundance were characterized by relatively narrow values of the vegetation index, but areas of low abundance showed a great deal of variability in the index (Fig. 4.5). This was true for both species. However, the plot for Baird's Sparrow had a single peak, corresponding

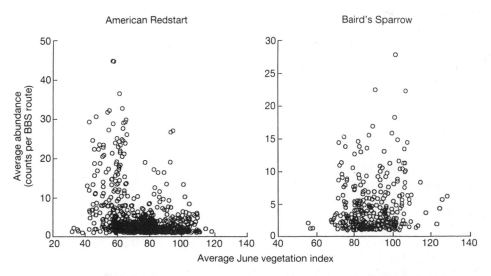

Fig. 4.5 Relationships between remotely-sensed vegetation data (an index based on reflectances of certain wave-lengths of light measured by a satellite orbiting the earth) and average abundances of American Redstart and Baird's Sparrow obtained from trend surfaces.

to the structure of its geographical range, while the plot for American Redstart had two peaks, each corresponding to a different centre of abundance. It is clear that although the plots correspond to well defined characteristics of the species' geographical ranges, there are some difficulties with using these relationships to develop linear regression models. Some of these difficulties may arise from the spatial auto-correlation present in the original patterns of abundance of the species, and may be resolvable using statistical models that explicitly assume spatial autocorrelation.

4.2 Measuring demographical fragmentation

Once estimates of the spatial patterns of density are obtained for a species, the next question that arises is how one goes about measuring demographical fragmentation. In the last chapter, I introduced the concept of a fractal dimension and how that quantity could be used to describe the shape of a spatial pattern. Measures of areographical fragmentation were based on counting the number of windows of a specific size needed to cover the outline of a species' range. Further insight into patterns of areographical fragmentation emerged from examining area–perimeter relationships. Demographical fragmentation is also an exercise in measuring the complex nature of a natural shape, but there are some complications. Consider for a moment that the spatial distribution of abundance can be conceived of as a topographical phenomenon, where the abundance peaks correspond to mountain ranges. The problem of measuring demographical fragmentation then becomes a problem of measuring how steep the abundance peaks are, and how broken-up the abundance landscape appears. Initially, one would think that it might be possible to use a three-dimensional window, or box, to cover the abundance pattern, then simply count the number of boxes it takes to include the pattern. However, this is not possible, because the height of abundance peaks are measured in different units than the distances between peaks. Hence, it is not possible to calculate a box dimension. In this section I consider alternative ways of examining demographical fragmentation using fractal geometry that circumvents the problem just posed.

4.2.1 Semivariograms, contours and fractal dimensions

The semivariogram provides a description of the spatial correlation between abundances at varying distances from one another. Burrough (1983) described theoretical analyses that indicate that patterns that arise from processes that are continuous, but not smooth, can be analysed using the semivariance. When the logarithm of the semivariance is plotted against the logarithm of lag distance, then the slope of this relationship is related to a fractal dimension that described how steeply peaked or 'crinkled' the surface is. Thus, a regression of log $g(h)$ against log h would give

$$\log g(h) = \beta_0 + \beta_1 \log h$$

The fractal dimension of the surface is then calculated as

$$D_s = 2 - \frac{\beta_1}{2} \tag{4.5}$$

The fractal dimension of the semivariance function is related to the amount of demographical fragmentation that exists across a species' geographical range. It describes the rate at which densities change from one place in space to the next. A species with high demographical fragmentation would show steep changes from one place to the next in its abundance, and consequently have a semivariance function with a high fractal dimension.

It is possible that the log−log semivariogram may not be perfectly linear, or may be 'piecewise' linear, that is, it may appear to be several straight lines put together (see Fig. 2.3). In this case, several processes operating at different scales may be responsible for generating the distribution of abundance. Recall that the semivariance function may have both a random and a deterministic component, and each of these may operate at different scales. Such a multiscaled process would show up in the semivariance having different slopes for different segments of the log−log semivariogram.

Pennycuick and Kline (1986) suggested another technique that can be modified to measure demographical fragmentation. It is based on measures of the areographical fragmentation of population abundance contours used to describe the surface. Suppose we want to obtain an estimate of the total number of individuals that would be counted if the entire geographical range of a species were censused with the same technique so that no part of the geographical range were uncensused. There are a number of ways to estimate this. Essentially, what is needed is an estimate of the volume underneath the abundance distribution. Assume that the abundance measurement is a density, and we want to sum all densities across space. Firstly, a collection of sites are sampled and densities obtained. Next, a surface of densities is estimated using kriging or some other interpolation technique. Since the abundance surface is always calculated using a grid with a predetermined grain size, we cannot just add the abundances we obtained at all points in the grid. Therefore, a contour map is obtained by choosing a density, d, and drawing contours starting at zero density at every d intervals. This gives a map that estimates the spatial distribution of population abundances such as the one shown in Fig. 4.2. Let the total area enclosed by contours of density level i be $A_i(h)$, where h is the window size, then an estimate of the total number of individuals across a species' geographical range is obtained by summing the areas of each contour multiplied by d. This is essentially an estimate of the volume contained beneath the density trend surface. Let $V(h, d)$ be this volume, then

$$V(h, d) = \sum_i d\, A_i(h)$$

Notice that the estimated volume under the surface depends both on the abundance interval used and the size of the two-dimensional window used to measure the area of each contour. The areographical fragmentation of each contour can be described using the box dimension of that contour. Using equation 3.7, the volume becomes

$$V(h, d) = d \sum_i a_i\, h^{2-D_{B_i}} \qquad (4.6)$$

where D_{B_i} is the box dimension of the ith contour. Now assume that the volume calculated in equation 4.6 is a fractal itself, so we could write

$$V(h, d) = v\left(\frac{1}{d}\right)^{D_v'}$$

Substituting this back into equation 4.6 gives

$$v\left(\frac{1}{d}\right)^{D_v} = \sum_i a_i\, h^{2-D_{B_i}} \qquad (4.7)$$

$$= A_s$$

where $D_v = D_v' + 1$. Taking logarithms gives

$$\log v + D_v \log\left(\frac{1}{d}\right) = \log \sum_i a_i\, h^{2-D_{B_i}} \qquad (4.8)$$

$$= \log A_s$$

The quantity A_s in equation 4.7 can be called the fractal area of the abundance surface. It depends on the density interval chosen, because when d is changed, the surface must be divided up into a new set of contour intervals. Equation 4.8 implies that demographical fragmentation can be measured by estimating D_v from a regression of the logarithm of the fractal area of the abundance surface on the logarithm of the inverse of the abundance interval. That is, given the regression equation

$$\log A_s = \beta_0 + \beta_1 \log\left(\frac{1}{d}\right) \tag{4.9a}$$

where

$$A_s = \sum_i a_i h^{2-D_{B_i}} \tag{4.9b}$$

An estimate of demographical fragmentation is given by the slope of the regression in equation 4.9a. The quantity D_v will be referred to as the 'volume dimension'.

4.2.2 Example analyses: demographical fragmentation

Maurer and Heywood (1993) used the semivariance function to estimate fractal dimensions for 85 species of birds. The data were obtained from the BBS. Here I discuss the results obtained for two species of congeneric wood warblers, Kentucky Warbler (*Oporonis formosus*) and MacGillivray's Warbler (*Oporonis tolemiei*) to illustrate in detail how fractal dimensions of the semivariance were obtained and the limitations of interpretation.

The plots of the logarithm of the semivariogram against the logarithm of lag distance indicated that the slope for MacGillivray's Warbler was more highly positive than for Kentucky Warbler (Fig. 4.6). Thus, using equation 4.5, the fractal dimension of the semivariance for MacGillivray's Warbler was 1.74 and that for Kentucky Warbler was 1.99. This corresponds well to the visual impression of the distribution of population abundances of the species obtained from kriging (Fig. 4.7). Kentucky Warbler has a distribution of abundances with multiple peaks and relatively steep (closely placed) abundance contours, indicating the abundance changes relatively rapidly across its geographical range for this species. MacGillivray's Warbler, in contrast, has a geographical range that extends a long way up the western coast of North America, and has a single major abundance peak with relatively widely spaced contours. A few smaller peaks, also gently rising, occur towards the southern portion of its range. Interestingly, both species have very similar degrees of areographical fragmentation: the geographical range of Kentucky Warbler has a box dimension of 1.69, while that of MacGillivray's Warbler has a box dimension of 1.64.

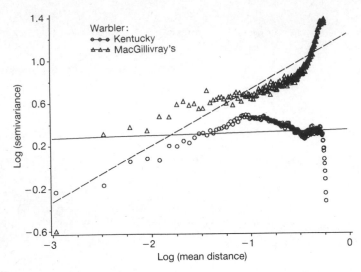

Fig. 4.6 Log–log semivariograms for two species of *Oporonis* wood warblers. The slope of each line is an estimate of the fractal dimension of the semivariogram, or semivariance dimension. Note that in each semivariogram there are significant nonlinearities, so the dimension for each species is only approximate.

It is important to point out that the log–log semivariograms in Fig. 4.6 are clearly not linear. The semivariogram for MacGillivray's Warbler seems to increase for large lags, while the semivariogram for Kentucky Warbler decreases. Such nonlinearities indicate that there may actually be processes operating at more than one scale (see section 2.1.3) for each species. The differences between species in their semivariograms may indicate that each species' geographical population is affected by factors that change in very different ways across space. It is important to note that these species live in very different places in North America. MacGillivray's Warbler lives in areas that are more topographically complex and climatically variable than Kentucky Warbler. It is not surprising that their semivariograms look different. The linear regressions' fit to each species may indicate something about these differences, but clearly they do not reflect the complexity of the situation. More research needs to be done to relate the properties of semivariograms of geographical populations to the processes which affect the patterns.

Maurer and Heywood (1993) interpreted these patterns of variation in semivariance dimensions as follows. For species of birds that migrate to the neotropics, the semivariance dimension was negatively related to average abundance: species with high degrees of demographical frag-

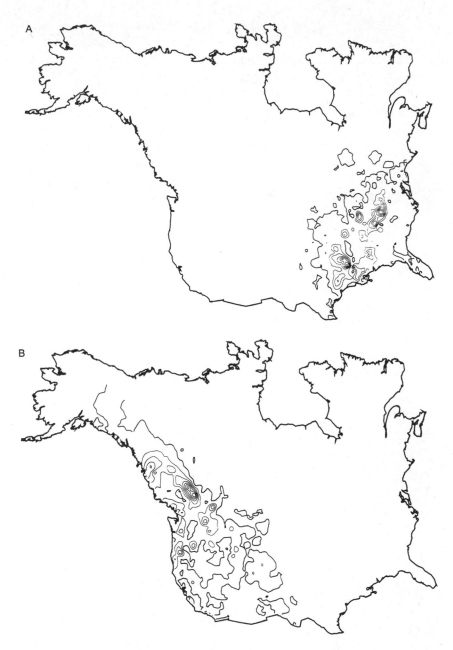

Fig. 4.7 Estimated geographical distribution of abundances for Kentucky Warblers (A) and MacGillivray's Warblers (B).

mentation tended to be rarer where they occurred on average. This was not true for birds that did not migrate to the neotropics. Maurer and Heywood (1993) hypothesized that this pattern was related to the sensitivity of neotropical migrants to changes in abundance. Note that for the two *Oporonis* warblers, the one that has a higher semivariance dimension (Kentucky Warbler) also tends to be less abundant where it is found.

Results such as Maurer and Heywood's (1993) analyses have important implications for conservation. If some species are more sensitive to population changes than others, then management efforts need to be more carefully implemented to maintain populations across species' geographical ranges. However, such patterns may only partially reflect the processes at work that shape species' abundance distributions. A closer examination of the semivariograms of the two wood warbler species portrayed in Fig. 4.6 indicates that other interpretations might be made. Firstly, note that the stationary portion of the semivariograms for both species are similar, but that the semivariogram for Kentucky Warbler has a considerably higher nugget variance. The nugget variance is thought to reflect the amount of variability within sampling points (e.g. Haining, 1990). Thus, the higher nugget variance for Kentucky Warbler implies that, on the average, census points for this species are more variable (temporally) than census points for MacGillivray's Warbler. Since temporal variability is thought to increase the likelihood of extinction for a species (e.g. Pimm, 1991), local populations of Kentucky Warblers would be more prone to extinction. Secondly, it seems likely that the drift portion of the semivariograms for the two species are very different. For MacGillivray's Warbler, the semivariogram continues to rise at long lag distances, while for Kentucky Warbler, there is a pronounced decline at long lag distances. This influences the estimate of demographical fragmentation for the species, and implies that perhaps more than one dimension should be considered for different parts of the semivariograms. Nevertheless, these differences in semivariograms support the contention that Kentucky Warbler populations are more severely limited than MacGillivray's Warbler populations.

Volume dimensions for each of 86 neotropical migrant birds and their relatives were calculated using equation 4.9. These are the same set of species used in the last chapter to analyse area−perimeter relationships. Volume dimensions were calculated by obtaining a series of sizes of the abundance window (d in equation 4.9). The maximum abundance on the map for each species was obtained. Next, the maximum abundance was divided by 1, 2, ... 8 to obtain 8 values for d.

The fractal area of the abundance surface was calculated for each of these values and its logarithm was plotted against log (l/d) to obtain an estimate of the volume dimension. Sample calculations for Kentucky and MacGillivray's Warblers are shown in Fig. 4.8. Recall that Kentucky Warbler had a higher semivariance dimension and that its geographical range is smaller than MacGillivray's Warbler. This implies that it has relatively steeper changes in abundance from one place to the next. It also has a steeper slope for the log fractal area of the abundance surface − log abundance window size relationship. This means that as one decreases window size (and thus obtains a closer approximation to the volume underneath the geographical abundance distribution) the fractal area of the surface increases relatively faster for Kentucky Warbler than for MacGillivray's Warbler. This is consistent with the interpretation that Kentucky Warbler has relatively steeper changes in abundance across its geographical range than MacGillivray's Warbler.

Plotting the semivariance dimension against the volume dimension for the 86 species of neotropical migrants and their relatives indicated that there was a significant positive correlation between the two measures of demographical fragmentation (Fig. 4.9). However, there was a fair degree of scatter about the relationship, and separate regressions for

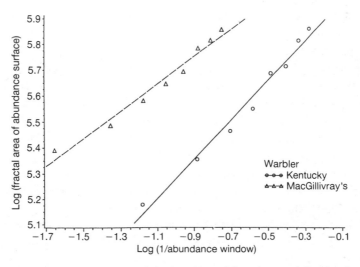

Fig. 4.8 Simple linear regressions of the logarithm of fractal area of the abundance surface (A_s, in equation 4.9) against log (l/d), where d is the size of the abundance window used to calculate A_s for two species of *Oporonis* wood warblers. Notice that the slope, and hence, the volume dimension, is higher for Kentucky Warbler than for MacGillivray's Warbler.

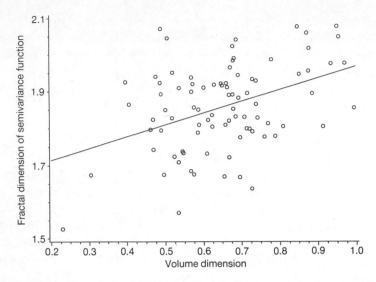

Fig. 4.9 Volume dimensions calculated for 85 species of neotropical migratory birds and related species that do not migrate to the neotropics plotted against semivariance dimensions. Note that species with higher semivariance dimensions tend to have higher volume dimensions. The correlation is significant ($r = 0.41$, d.f. $= 84$, $P < 0.01$). The line was obtained by ordinary least squares.

each group of migrants were not significantly different. I interpret this to mean that both measures are related to different aspects of demographical fragmentation. The semivariance dimension is closely related to the patterns of spatial covariation of abundance. The positive relationship of the semivariance dimension with average abundance shown by Maurer and Heywood (1993) implies that covariation in abundance is relatively higher for widespread and abundant species than for rarer species. The volume dimension, on the other hand, looks at the spatial continuity of abundance contours. That is, it measures how spatially fragmented areas of high abundance are. The relationship between the semivariance and volume dimensions suggests that high degrees of spatial autocorrelation in abundance are related to more contiguous spatial patterns of areas of high abundance. That is, the central populations of widespread and common species are less fragmented than those of rarer species.

4.3 Summary
Spatial patterns of geographical populations can be analysed using a number of techniques. I have emphasized here techniques based on

geostatistical models that incorporate the GVRM. There are a number of technical problems with using the semivariogram to develop methods of interpolation. In particular, it does not appear that spatial patterns in geographical populations are stationary. From a theoretical viewpoint, populations should be high in the centre of their geographical range, although there may be more than one region of high abundance. This means that the drift component of the semivariance will contain significant information about the spatial pattern of a particular species. As was seen in the example analyses, this component complicates interpretation of the semivariance dimension as a measure of demographical fragmentation. Although it was calculated as a single number in the examples, there are probably several semivariance dimensions for each species, each referring to a set of processes that manifest themselves at different scales. Systematic study of the properties of the semivariance of geographical population abundance should provide further insight into geographical patterns of population abundance.

Spatial patterns in population abundance using extensive population surveys suffer from some potential biases. In the example analysis presented here, the relationship between extensive and intensive censuses appeared to be nonlinear and subject to a large amount of sampling variation. The nonlinearities were not removed by some simple transformation such as taking logarithms. Instead, there appeared to be a region where extensive censuses were relatively 'insensitive' to changes in abundance of intensive censuses.

Abundances from extensive censuses can be correlated with variation in environmental conditions, such as vegetation density. Such data can be obtained from remotely sensed data, such as satellite images. Initial analyses of single measures of vegetation density in June, the season when BBS routes are run, indicated that sites with low abundances tended to be associated with a wide range of vegetation densities. Sites where species had the highest abundance occurred in a relatively narrow range of vegetation densities.

Demographical fragmentation can be measured as the degree of steepness or brokenness of a surface of population abundances. Fractal dimensions associated with such surfaces can be obtained from at least two techniques. Firstly, logarithmic transformation of the semivariance provides an estimate of the semivariance dimension. As indicated above, the interpretation of this statistic is complicated by multiple scales of variation in abundances. A different approach is based on the use of contours of various sizes. The rate at which the surface area of the abundance distribution changes with changes in the 'thickness' of the intervals separating contours gives another fractal measure of demographical fragmentation.

Clearly, there is a great deal that remains to be done in the analysis of spatial patterns of abundance across geographical ranges. From the pioneering work of L.R. Taylor to modern advances in computing made available by GIS technology, the subject of geographical variation in abundance remains one of the most challenging and potentially interesting provinces of biogeography (Hengeveld, 1990). Theoretical advances in understanding how spatial patterns of populations across space lead to sensitivity to environmental changes will await further progress in identifying data sources and statistical techniques to describe these patterns. More importantly, practical approaches to identifying species that are most likely to be sensitive to changing global ecosystems are likely to gain appreciably from extensive analyses of the spatial patterns of geographical populations.

CHAPTER 5

Geographical population dynamics

In the preceding chapters, geographical patterns of population abundance have been analysed as if the patterns were static. This approach might be suitable to answer some questions, but neither range boundaries nor spatial patterns of population abundance remain constant over time (Taylor, 1986; Hengeveld, 1990). Detailed analysis of the temporal dynamics of geographical populations requires long-term data across many locations, and therefore is more severely limited by available data than the analyses previously discussed. However, to address questions about the persistence of a species through time, techniques to analyse temporal dynamics of geographical populations are needed. In this chapter, I address the question of how one would go about analysing temporal changes in geographical patterns of abundance.

One way to approach the temporal aspect of geographical abundance patterns is to partition a data set into time periods, and perform the types of analyses described in the previous chapters separately for each time period. Thus, one could calculate measures of areographical and demographical fragmentation from a temporally partitioned data set, and observe the pattern of changes in these measures over time. One could, for example, ask the question, 'Does demographical fragmentation increase over time for a particular kind of species?' For example, in Fig. 5.1, the average amount of demographical fragmentation for neotropical migrant and resident bird species is shown for five time periods. These were calculated by dividing BBS censuses into 5-year blocks (see discussion in section 5.1.1), and estimating the semivariance function for each species during each time period. Interestingly, the data indicate that demographical fragmentation has increased on the average during the past 25 years.

Analyses such as those described in the last paragraph can illuminate many aspects of how geographical populations change over time. In this chapter, I will describe more direct techniques that emphasize temporal patterns in geographical populations. These methods all assume that geographical populations change due to a large number of environmental changes across a species' geographical range, and therefore, the dynamics observed should have a very complex set of causal mechanisms if one were able to identify all of the individual events that together culminate in changes in geographical populations.

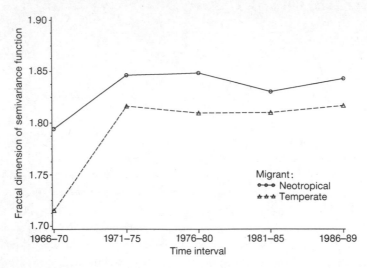

Fig. 5.1 Changes in average degree of demographical fragmentation (measured by the semivariance dimension) over time for two groups of passerine birds. The increase in fragmentation over time in both groups was significant, but group means were not significantly different.

5.1 Changes in spatial trend surfaces over time

In the last chapter, I discussed methods of describing the spatial distribution of abundance taking advantage of spatial autocorrelation of abundance among sampled sites. If a large number of sites have been censused over a relatively long period of time, then it is possible to divide the data set up into evenly spaced time periods providing there are a sufficient number of accurately sampled sites during each time period to allow estimation of a trend surface. The trend surface for each time period can then be analysed as if it were part of a time series of trend surfaces. This method was first used by L.R. Taylor to describe the dynamics of insect distributions in Great Britain (Taylor, 1986).

Let $X(t)$ be a matrix that contains the population abundances of a species in cells of a regularly spaced grid across geographical space. In GIS, data stored in such a grid is called a 'raster image' or 'representation' (Burrough, 1986). Each individual point in the matrix, which we will denote as $X(t)_{ij}$, gives the abundance of the species during time interval t at the i,jth point on the grid. The average rate of population change at each point on the grid between time t-1 and time t is calculated as

$$r(t)_{ij} = \ln \frac{X(t)_{ij}}{X(t-1)_{ij}} \tag{5.1}$$

Thus, a matrix of 'average rates of population change', $r(t)$, can be constructed for each time period that contains the individual $r(t)_{ij}$s. This matrix can also be stored as a raster image in a GIS for further analysis.

5.1.1 Example analyses: geographical population dynamics of Brown-headed Cowbirds and Cerulean Warblers

Two species of North American birds have been of particular concern to conservation biologists for different reasons. Cerulean Warblers (*Dendroica cerulea*) have declined significantly in the past 25 years in North America (Robbins *et al.*, 1992). They appear to be experiencing significant losses of habitat on both their wintering and breeding grounds. Brown-headed Cowbirds (*Molothrus ater*) are nest parasites on many species of birds, and they have been implicated as an important factor in the local declines of a number of species (e.g. Robinson, 1992). I used these two species because they represent two extremes. The Cerulean Warbler is narrowly distributed and occurs at relatively low abundances (average counts per BBS route, 2.6). It has a relatively high degree of demographical (D_S, 2.07) and areographical (D_B, 1.74) fragmentation. Furthermore, it is a neotropical migrant species, suggesting that it is more sensitive to environmental changes than species that do not migrate to the neotropics (Maurer & Heywood, 1993; Heywood, 1992). Brown-headed Cowbirds represent the other extreme. They are widely dispersed and relatively common (average counts per BBS route, 15.8), have relatively low degrees of demographical (D_S, 1.91) and areographical (D_B, 1.86) fragmentation (recall that areographical fragmentation decreases with increasing box dimension), and do not migrate to the neotropics.

For each of these species, a trend surface was estimated by universal kriging using BBS data for each of following 5-year periods: 1966–1970, 1971–1975, 1976–1980, 1981–1985 and 1986–1989. Here I will discuss only the changes that occurred in the species' abundance patterns between 1976–1980 and 1981–1985. However, the patterns for these years were similar to those for other years.

Cerulean Warblers are most common in the southeastern United States in the southern Appalachian Mountains (Fig. 5.2). They are much less common further to the north and west of this region of high abundance. As noted above, they tend to be relatively rare on the average. The maximum abundance that they achieve, however, is around 12 birds per BBS route. The Cerulean Warbler is a good example of a rare, restricted species with relatively restricted habitat

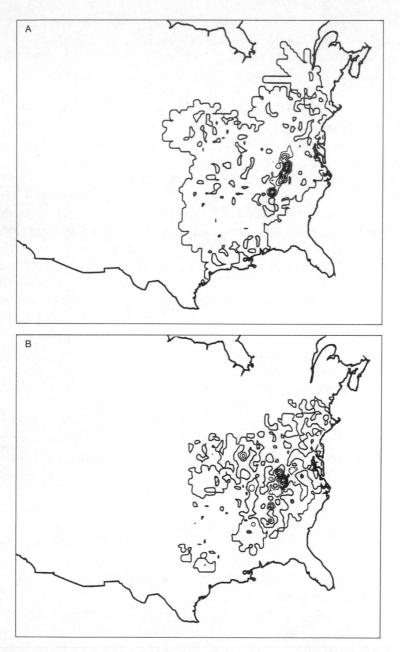

Fig. 5.2 (A) Geographical distribution of abundance for Cerulean Warblers obtained by universal kriging. Data on abundance were obtained from the BBS. The maximum abundance was around 14 birds per BBS route. (B) Relative changes in abundance of Cerulean Warblers from 1976–1980 and 1981–1985. Contours with hatching indicate regions of population decline; those without hatching indicate regions of population increase.

requirements (Robbins *et al.*, 1992), and hence, relatively restricted geographical distribution. The surface representing relative changes in abundance between 1976–1980 and 1981–1985 indicated that the steepest changes occurred in those regions where the species was most abundant. In a few regions, those changes represented increases, but decreases were more widespread.

Robbins *et al.* (1992) indicated that Cerulean Warblers have consistently decreased in abundance since the 1960s (Fig. 5.3). They calculated abundance as the average of counts obtained each year on a sample of 255 BBS routes. I recalculated abundances as the average count each year of only those routes that had non-zero counts. I found that for these routes, Cerulean Warblers decreased between 1980 and 1985, but increased again after 1985 (Fig. 5.3). The difference between my results and those of Robbins *et al.* (1992) is important to contrast here, because it says something about the mechanisms of geographical population changes. Robbins *et al.* included zero counts in their averages, while I did not. The steady decline evident in their data is only partially attributable to local changes in abundance. Their yearly estimates reflect an increasing number of routes on which zero counts for Cerulean Warblers were obtained. Therefore, the decline that they documented is mainly due to a shrinkage of the species' geographical range. There did appear to be an event lasting nearly 5 years that was responsible for declines within local populations, and this may have been responsible for the loss of many local populations near the periphery of the geographical range. Nevertheless, the greatest magnitude of changes occurred within the centre of the species' geographical range. Even when these central populations recovered, the effects of geographical range shrinkage were seen in the declining average abundances in the data of Robbins *et al.*

Brown-headed Cowbirds are widespread throughout North America (Fig. 5.4), much to the dismay of many conservationists intent on maintaining healthy populations of neotropical migratory birds. As mentioned above they are relatively common wherever they are found, and reach spectacularly high maximum abundances (around 190 counts per BBS route). In contrast with the narrowly distributed Cerulean Warbler, relative changes in abundances of Brown-headed Cowbirds were not concentrated in the centre of their range, but distributed widely across those parts of the continent where they were found (Fig. 5.4).

It is not clear whether these differences in the distribution of changes in abundance between a relatively rare and localized species

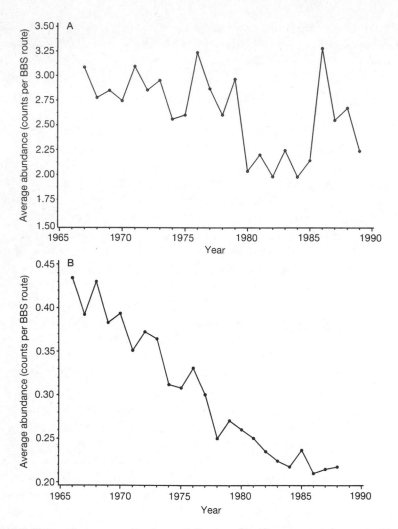

Fig. 5.3 Changes in average abundance of Cerulean Warblers across their geographical ranges as calculated by myself (A) and by Robbins *et al.* (1992) (B). The discrepancy between the patterns is due to the fact that Robbins *et al.* calculated the means using routes with zero counts and I did not. The decline of Cerulean Warblers was due not only to a drop in abundance in areas where they were found (which actually returned to previous levels after 1985), but also to a decrease in the number of sites occupied (i.e. shrinkage of the geographical range).

and a common and widespread species are typical of all species. This would require examination of literally hundreds of different maps of changes in abundance. Such an exercise might be fruitful, but would clearly be a long-term task. One problem is that there are no objective measures of how concentrated these positive and negative changes in

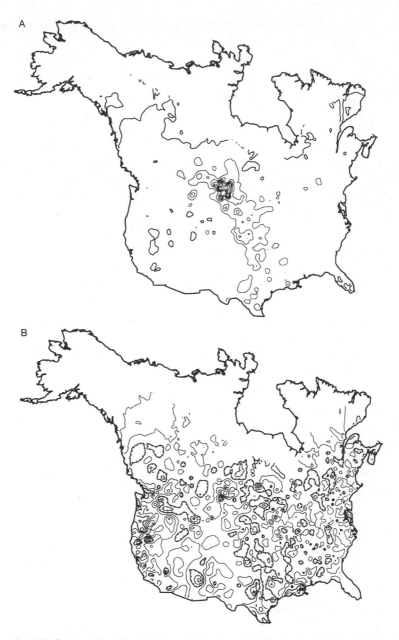

Fig. 5.4 (A) Geographical distribution of abundance for Brown-headed Cowbirds obtained by universal kriging. Data on abundance were obtained from the BBS. The maximum abundance was over 100 birds per BBS route. (B) Relative changes in abundance of Brown-headed Cowbirds from 1976–1980 and 1981–1985. Contours with hatching indicate regions of population decline; those without hatching indicate regions of population increase.

abundance are. The analyses presented here are certainly suggestive of some profound differences in the geographical population dynamics of rare and common species. In the next section, I consider how analysis of spatial and temporal variability in population abundance may shed further light on this problem.

5.2 Spatial and temporal variation in abundance

Interest in patterns of variability in population abundance was stimulated by observations that some species tend to vary more than others. On the one hand, some species of insects, for example, often undergo spectacular irruptions that can be economically very important. On the other hand, many species of vertebrates seem to maintain relatively constant levels of abundance. All species, however, show some level of variability in population abundance. Pimm (1991) has reviewed studies which indicate that population variability is related to the ability of a population to persist. Theoretical models of stochastic population dynamics suggest that demographical and environmental variability can have a profound negative effect on the ability of a population to maintain itself (e.g. Leigh, 1981; Goodman, 1987a,b). It does not necessarily follow that populations that show high degrees of variation in abundance are necessarily experiencing high degrees of demographical and environmental variability. Nevertheless, there are many empirical studies that show that populations that have high degrees of variation in abundance also tend to have higher likelihoods of becoming extinct (see Pimm, 1991 for a review of these studies). Of interest to the present discussion are spatial patterns in temporal variability across a species' geographical range. In this section I consider methods that can be used to study these patterns.

5.2.1 Population variability: Taylor's power law

In a pioneering paper, L.R. Taylor (1961) found that for a number of different populations of invertebrate animals, the variance in population abundance could be expressed as a power function of average abundance, so

$$s^2 = a'x^{b'}$$

where a' and b' are constants, x is the average abundance, and s^2 is the variance. The standard deviation is then given by

$$s = ax^b$$

where $b = b'/2$ and $a = (a')^{1/2}$, and the coefficient of variation of abundance is

$$cv = ax^{b-1} \tag{5.2}$$

The means and variances can be calculated either from a single population sampled over a number of years, in which case one needs a number of populations at different points in space to estimate the parameters a and b, or they can be calculated from a set of populations sampled at the same time, in which case one would need collections of populations from several years to estimate a and b (Hanski, 1982). The parameters obtained from each of these two situations in general will not be equal, and should be interpreted differently.

The parameter b' has traditionally been interpreted as a measure of aggregation (Taylor, 1961; Taylor & Taylor 1977; Hanski, 1982). Since a Poisson distribution has a mean equal to its variance, if $b' = 1$, the implication is that populations are randomly distributed in space or time (depending on how b' is estimated). However, equation 5.2 suggests a slightly different interpretation. If a collection of sites across the geographical range of a species were sampled over time, and b was estimated from temporal means and variances, then b is a measure of the relative amount of variability in places where the species is abundant compared to places where the species is rare. Since a species can have one or more centres of abundance across its geographical range, b compares the relative variability of populations in these centres of abundance to those at the range periphery. Thus, if $b < 1$, then the coefficient of variability of central populations is lower than for peripheral populations, that is, peripheral populations are relatively more variable than central populations. If $b > 1$, then the opposite is true, that is, central populations are relatively more variable than peripheral populations.

A similar interpretation can be made if calculations are done with spatial means and variances instead of temporal ones. That is, for each year, censuses are averaged over space, and the parameters in equation 5.2 are calculated using years as replications. In this case, if $b < 1$, then there would be relatively less spatial variation in years when the species was less common than when it was abundant. If $b > 1$, then there would be relatively more spatial variation in years when the species was abundant.

5.2.2 Example analyses: spatial and temporal variability
Spatial and temporal variability across species' geographical ranges were estimated for 85 species of North American birds in four taxa as follows (the same taxa as was used in sections 3.2.3 and 4.2.2). The BBS data set was used to extract all censuses on which a species was

recorded over 23 years of censuses. To obtain an estimate of the spatial pattern of temporal variability, each census was classified into one of 10 abundance classes based on the proportion of the maximum abundance that the average on that census represented. Abundance classes were scaled logarithmically because abundances per census followed a 'hollow curve' distribution, i.e. there were many censuses with low abundances and only a few with high abundances. The logarithmic abundance classes tended to even out the number of censuses among classes. Within each abundance class, the mean and variance were calculated across years for all censuses that fell in that abundance class. The means and variances were then used to estimate the parameter *b*. The logarithm of the standard deviation was plotted against the logarithm of the mean for each abundance class, and the slope of the regression line was calculated by ordinary least squares. Recall that if this slope is less than 1, then peripheral (least abundant) populations vary more over time than central populations. If the slope is greater than 1, the opposite is true.

A similar procedure was used to examine temporal patterns of spatial variation. For every year, the mean and variance in abundance were calculated across all BBS census routes on which a species was observed. The logarithm of the standard deviation was plotted against the logarithm of the mean for each species. Slopes for this regression that were less than 1 indicated that in years when the species was more abundant, spatial variability was relatively less than when the species was less abundant. A slope greater than 1 indicated that a species was more spatially variable when it was common than when it was rare.

Examples of regressions of log standard deviation against log mean abundance are given for Cerulean Warblers and Brown-headed Cowbirds in Figs 5.5 and 5.6, respectively. In the last section we saw that changes in Cerulean Warbler populations were concentrated in the centre of their geographical ranges. The slope describing the spatial pattern of temporal variation for this species was 1.55, considerably greater than 1. Likewise, the slope for the temporal pattern of spatial variation was 1.51. Thus, Cerulean Warblers tended to have central populations that were much more variable than peripheral populations, confirming the analysis in Fig. 5.2. Furthermore, Cerulean Warblers tend to have a relatively higher degree of spatial variability among populations when they are common than when they are rare. In contrast to Cerulean Warblers, Brown-headed Cowbirds tended not to have changes in abundance concentrated in the centre of their geographical ranges (Fig. 5.4). The slope describing the spatial pattern of

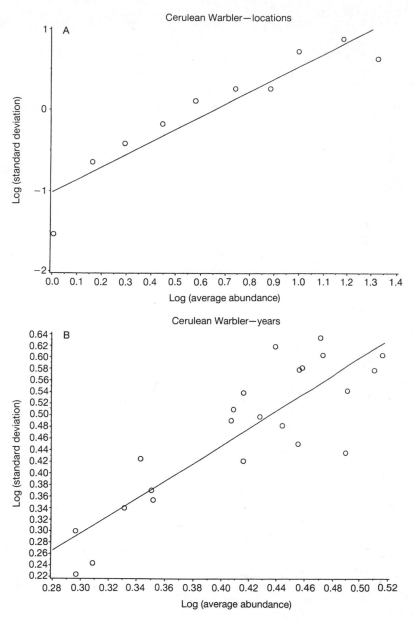

Fig. 5.5 Regressions of the logarithm of the standard deviation of counts against the logarithm of average abundance for Cerulean Warblers. (A) represents temporal means and standard deviations calculated within 10 logarithmic abundance classes. (B) represents spatial means and standard deviations calculated for each of 23 years.

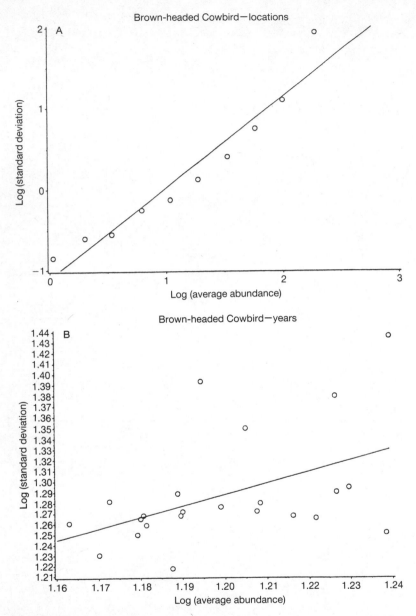

Fig. 5.6 Regressions of the logarithm of the standard deviation of counts against the logarithm of average abundance for Brown-headed Cowbirds. (A) represents temporal means and standard deviations calculated within 10 logarithmic abundance classes. (B) represents spatial means and standard deviations calculated for each of 23 years.

temporal variation for the Cowbird was 1.12, not too different from 1. In addition, the slope for the temporal pattern of spatial variation was 1.07, again very close to 1. Therefore, Brown-headed Cowbirds have central populations that vary about the same degree as peripheral populations, and do not seem to vary any more in years when they are more abundant than when they are less so. Again, this accords well with the patterns seen in Fig. 5.4.

For the 85 species of neotropical migrants and their relatives, it was found that there was a positive relationship between the slopes for the spatial pattern of temporal variation and the temporal pattern of spatial variation (Fig. 5.7). That is, species which have central populations that vary more than peripheral populations tend to have a higher degree of spatial variation when they are common than when they are rare. Interestingly, the relationships between the spatial and temporal aspects of variation for neotropical migrants and residents were different (Fig. 5.7). For a given degree to which a species is more variable when it is common than when it is rare, a neotropical migrant will have relatively more variable central populations than a resident. This implies

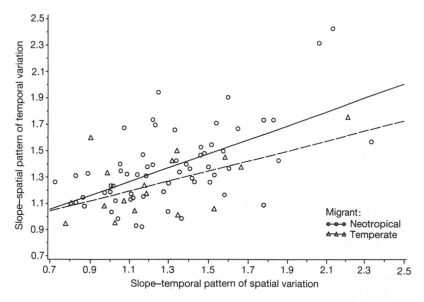

Fig. 5.7 Relationships between the slope obtained from regressions of temporal means and standard deviations and the slope obtained for regressions of spatial means and standard deviations for 85 species of neotropical migratory birds and their relatives that do not migrate to the tropics. Notice that the slope of this relationship is positive for both groups and is higher for neotropical migrants.

that the central populations of neotropical migrants tend to be less stable than those of residents, if variability is related to persistence (Pimm, 1991).

A problem with the analyses reported here is that it was very difficult to determine how to handle zero abundances. This is because BBS routes are often run only irregularly, so that one year, on a census within a species' geographical range, a zero abundance might be obtained for that species, and in another year, the route was not run, so the abundance is unknown for that year. This makes it difficult to interpret zero abundances. In the analyses listed above, I left out all zeros, but this may have underestimated the degree of variability, particularly near the edge of the geographical range. Recently, S. Pimm (personal communication) has looked at this problem, and has made progress in developing methods to incorporate data on zero abundances. The tentative conclusions reached here may be modified as his method is used to analyse data like those presented here.

5.3 Geographical populations as nonlinear systems

Biological systems are among the most complex systems known to science. Geographical populations are no exception. As I indicated earlier in this book, geographical populations are complexly organized, with various levels of organization nested within other levels. Furthermore, the components that make up geographical populations (e.g. individual organisms, local populations, metapopulations) each participate in a number of other kinds of hierarchies, such as ecosystems and phylogenetic lineages. It is therefore clear that geographical populations must be influenced by many different processes as they change through space and time.

Recently, a great deal of interest has been focused on how to describe the behaviour of complex systems. In the last two decades, researchers have identified a kind of behaviour of a changing system that originates from mathematically well defined systems of differential equations, but shows irregular, aperiodic fluctuations. These models are often associated with what are called 'nonlinear' processes. This means that the differential equations used to describe the process include nonlinear terms, that is, the variables in the equations are raised to powers different from one or multiplied times one another. Such systems of equations generate solutions that have been misnamed 'chaos' because the solutions produce irregular fluctuations. Some of these models have been applied to biological populations, with limited success (Schaffer, 1984, 1985; Schaffer & Kot, 1985; Pimm, 1991). Since time series data obtained from censuses of biological populations often appear to show the same kind of qualitative behaviour, it is intriguing to ask if biological populations can be treated using the same kinds of techniques that are used to analyse the mathematical models that show irregular fluctuations. Generally, the answer appears to be that most data sets obtained from biological populations have far too few observations to apply the techniques commonly applied to analyse nonlinear, chaotic models.

In this section I describe two different kinds of nonlinear dynamics that originate in two fundamentally different ways. I then describe a probabilistic approach to analysing the time series produced by such models and suggest how the different kinds of nonlinear systems can be qualitatively differentiated using the approach. Finally, I describe how the technique can be applied to GPA using as examples the geographical populations of two species of wood warblers censused by the BBS.

5.3.1 Dynamics of nonlinear systems: 'chaos' and constrained random motion

There are probably any number of ways that a nonlinear system can be put together. Here, I will describe two fundamentally different kinds of nonlinear systems (out of many possible ones). The first kind is completely deterministic. Such sytems generally can be specified by as little as a single difference equation, or three differential equations. These equations always have nonlinear terms in them. A linear equation is made up of the sum of one or more variables, each of which can be multiplied by a constant. Such equations are called linear because the equation for a straight line is written in this way. A 'system' of linear differential equations can be written as

$$\frac{dX_i}{dt} = \sum_{j=1}^{k} a_{ij} X_j$$

where k is the number of variables (and differential equations) in the system and the a_{ij}'s are constants. The behaviour of such systems of equations is well understood, and they were used by Lotka (1925) to build a theoretical approach that was later incorporated into community ecology and led to some fundamental predictions about how communities should behave (see MacArthur, 1972). Systems of nonlinear differential equations can have quite different behaviour. One of the simplest and best known of these systems was first used by Lorenz (1963) to describe the essence of turbulent flows in the atmosphere. They are given as

$$\frac{dx}{dt} = a(y - x)$$

$$\frac{dy}{dt} = bx - y - xz$$

$$\frac{dz}{dt} = xy - cz$$

where a, b and c are constants. The nonlinearities arise in the second and third equations, where the variables x and z and the variables x and y, respectively, are multiplied by one another.

Time series obtained from a nonlinear system such as Lorenz's have an odd type of behaviour. Instead of changing smoothly from one time to the next, they show erratic changes, and one's ability to accurately predict those changes drops off rapidly with time. More interestingly, when one begins the time series at slightly different values, the two trajectories begin to diverge until it seems that the time series share nothing in common at all (Fig. 5.8). It is this property of such systems of equations that gives rise to the label 'chaos'.

The second kind of nonlinear system can arise by a very different kind of process, which I call constrained random motion. Consider an animal moving across an open field. The animal cannot be just anywhere in the field at any given point in time. Since it has a finite speed and ability to change direction, the animal's position in the field at any point in time is constrained by its position in the previous instant. Its position in the next instant will be constrained by where it is now. It may be wandering aimlessly with no particular direction in mind, and *eventually* it may move about across the whole field. But its motion is still constrained by its finite speed and ability to change direction. Thus, its motion is random, but constrained by its previous history of movement. The longer the time between instances when we note the position of the animal in the field, the less constraint, or correlation, we are likely to see between positions. The same kind of thinking can be applied to population dynamics. For example, an isolated population cannot be composed of two birds in one instant and two million birds in the next (unless, of course, each individual has a *very high* fecundity and development to the adult stage is instantaneous). For virtually any species, clutch sizes are finite, and development into adults takes time.

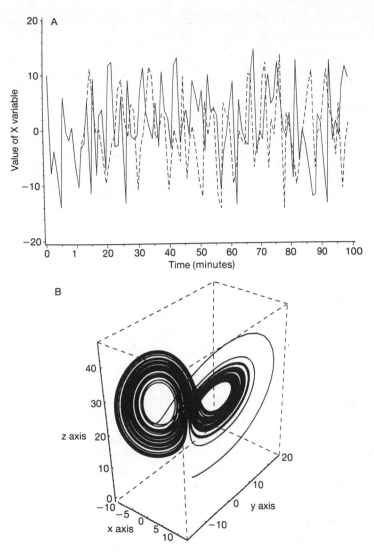

Fig. 5.8 (A) The time series generated by the Lorenz system of nonlinear differential equations for the first variable (x in the equations given in the text). The two time series shown were started from points that differed only very slightly. Notice that the trajectories stay within the same general bounds, but become increasingly different as time increases. (B) One solution to the Lorenz system of equations plotted in the three-dimensional phase space described by the differential equations. Notice that the solution stays on a very well defined region of the state space.

A population of two birds may eventually become a population of two million, but the population of two is more likely to be a population of 10 or 100 before it becomes a population of two million. Population change may be entirely random, but the direction and rate of change will generally depend on how many individuals were present in the population at the previous instant in time.

One way to model constrained random motion is to let the probability distribution depend in some way on the previous values that the state variable takes on. Consider the following two simple models. For the first model, at each point in time, we draw a value from a normal probability distribution with a mean of 0 and variance σ^2. Each draw from the normal distribution is independent of the next. Such a time series, shown in Fig. 5.9, is called 'Gaussian white noise'. The motion described by this model is random, and unconstrained. That is, the value at any given point in time has no statistical relationship with any other points in the time series, because each successive point is an independent draw from a probability distribution. For the second model, at each point in time, a draw is taken from a normal probability distribution with variance σ^2, but with a mean equal the value drawn in the previous time step (when $t = 0$, the mean is equal to zero). Thus, the mean of the distribution changes each time, and the random draw at each time step will be close to the one from the previous time step. A time series constructed in this way is also illustrated in Fig. 5.9. This second model is called 'Gaussian brown noise' or fractal Brownian motion' (see, for example, Voss, 1988). Fractal Brownian motion is a form of constrained random motion, where the value of the function at any given point in time depends on the previous history of values realized by the system.

Geographical populations could be governed by either kind of non-linear model described above. If the factors that govern the spatial and temporal dynamics of a geographical population are primarily deterministic, then the best model for geographical populations may be provided by a set of nonlinear differential or difference equations. If, however, there is a large component of chance in the dynamics of a geographical population, then a model of constrained random motion may be more appropriate. There is an intriguing third possibility. There may be so many deterministic processes that influence the dynamics of a geographical population, that the dynamics might be modelled by a constrained random motion model despite the entirely deterministic nature of the processes. Such a model would essentially be a statistical mechanical model of the geographical population in the

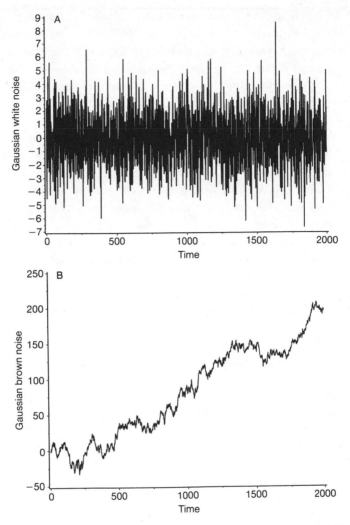

Fig. 5.9 Two time series generated by a normal probability distribution. (A) The time series was obtained by successive independent draws from a normal distribution. (B) The time series was generated by using the value obtained by a draw from a normal probability distribution as the mean of the next draw.

same sense that the pressure of an ideal gas can be modelled as macroscopic phenomena even though it results from a large number of deterministic events involving individual molecules. This kind of statistical mechanics for biological populations was first envisioned by Lotka (1925).

5.3.2 Calculating the information dimension of a time series

The two types of nonlinear models described in the last section could each apply to geographical populations to explain their temporal and spatial dynamics. In this section, I develop a framework for differentiating between these models that incorporates recent statistical ideas from the study of nonlinear systems.

A general concept that has emerged from the study of nonlinear systems is that nonlinear dynamical processes leave a 'signature' in the statistical distribution of values obtained from a time series generated by the system (see Ruelle, 1989 and Lasota & Mackey, 1994 for technical details). One way to read that signature is to calculate various 'dimensions' that describe the time series generated by the system (see Farmer et al., 1983; Parker & Chua, 1989). One dimensional measurement that seems to be particularly useful for the present situation is called the 'information dimension'. The information dimension is defined as follows. Suppose that we have a chaotic attractor such as the one shown in Fig. 5.8. Such an attractor generates a time series of values for each variable. The information dimension of the attractor is calculated using the statistical distribution of values realized by the time series generated by the attractor. To calculate this dimension, the range of possible values that the time series can take on is divided into small, discrete intervals, each of size h. If $n_i(h)$ is the number of times the time series has values that fall in the ith interval of size h, then the probability of falling in that interval can be calculated as

$$p_i(h) = \frac{n_i(h)}{N}$$

If T is the number of intervals where there are values for the time series, then

$$N = \sum_{i=1}^{T} n_i(h)$$

That is, N is the total number of observations in the time series.

Next, the statistical entropy of the time series is calculated as

$$H(h) = - \sum_{i=1}^{T} p_i(h) \log p_i(h) \tag{5.3}$$

where the base for the logarithms is arbitrary (I use base 10 in the calculations below). The statistical entropy of the time series can be partitioned as

$$H(h) = \log N - I(h) \tag{5.4a}$$

where

$$I(h) = \sum_{i=1}^{T} p_i(h) \log n_i(h)$$

$$= \log N - H(h) \tag{5.4b}$$

The quantity $I(h)$ is called the statistical information of the time series. From equation Fig. 5.4b it can be shown that as more observations are accumulated in the time series, then the statistical information increases (Maurer, in press). When each h-interval has only a single observation in it from the time series, then the statistical information is zero. When observations are equally distributed among the h-intervals, then the statistical information is equal to log N. An example of how to calculate the statistical information of a time series is given in Fig. 5.10 for a short time series.

The information dimension, D_I, can be defined as

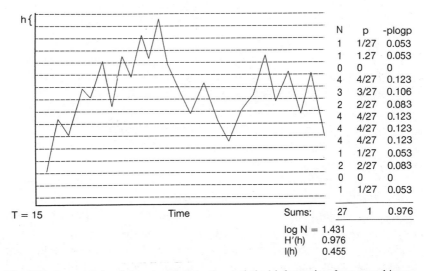

	N	p	-plogp
	1	1/27	0.053
	1	1.27	0.053
	0	0	0
	4	4/27	0.123
	3	3/27	0.106
	2	2/27	0.083
	4	4/27	0.123
	4	4/27	0.123
	4	4/27	0.123
	1	1/27	0.053
	2	2/27	0.083
	0	0	0
	1	1/27	0.053
Sums:	27	1	0.976

T = 15   Time

log N = 1.431
H'(h)   0.976
I(h)   0.455

Fig. 5.10 An example of how to calculate the statistical information from an arbitrary time series in one dimension. The abscissa represents time and the ordinate is divided up into arbitrary intervals of size h. The number of times values of the time series fall into each of these intervals is counted, and the probability of falling in each interval is calculated by dividing this number by the total number of observations (in this case $N = 27$). The statistical information is then calculated using equations 5.3 and 5.4. The information dimension would then be estimated by doing the same analysis for several different values of h.

$$D_I = \frac{\partial H(h)}{\partial \log (1/h)}$$

$$= \frac{\partial [\log N - I(h)]}{- \partial \log h} \tag{5.5}$$

$$= \frac{\partial I(h)}{\partial \log h}$$

That is, the information dimension is the slope of the relationship between the statistical information and the size of the h-interval used to calculate it. If this relationship is linear, then the slope of a simple linear regression of statistical information on $\log h$ will give an estimate of the information dimension.

What does the information dimension tell us about a time series? To answer this question, the information dimension was calculated for several time series. A three-dimensional time series of 2000 observations was generated using the Lorenz system of equations given above and the information dimension calculated as described above. Since there are three variables in this time series, the maximum value for the information dimension would be three. Since the attractor does not fill the space entirely, its information dimension is less than three. Information dimensions were also calculated for time series of Gaussian white and brown noise. The former represents unconstrained random motion, the latter, constrained random motion. The value obtained for information dimension depends on the length of the time series, so information dimensions were calculated for each time series for the first quarter of the time series, the first half, the first three-quarters, and the full time series. Differences in the way the information dimension changed with time for each of the model time series was evident from these calculations.

For all calculations, statistical information increased over time. This general dynamical result confirms Brooks and Wiley's (1988) speculations about the dynamics of information in evolution. However, the interpretation is quite a bit different from what they had in mind. Here, we do not assume that statistical entropy and information have any physical meaning other than as an accounting device for changes in the state of the system described by the time series. Plots of the cumulative statistical information at various times against the log of the measurement scale (h in equation 5.4), show how statistical information is related to measurement scale for the time series derived from the Lorenz system and the Gaussian noise systems (Figs 5.11, 5.12). Notice that none of these plots are linear over the entire range of measurement

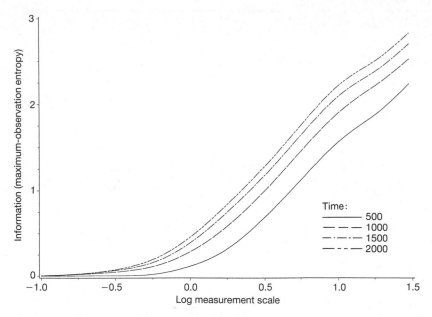

Fig. 5.11 Statistical information as a function of the logarithm of measurement scale (h) and time (different points in the time series are represented by different curves) for the Lorenz attractor. Note that this relationship is only linear for a relatively small range of measurement scales, but that this range increases with time.

scales. Within the range of measurement scales, however, there is a window of scales over which the statistical information is nearly a linear function of the measurement scale. Thus, for these systems, there is a 'natural' measurement scale over which the information dimension can be calculated as a linear relationship. The presence of this window of measurement scales is due to the finite length of the time series (Maurer, in press). If one had an infinite time series at one's disposal, then the statistical information − log measurement scale relationship would approach linearity across all measurement scales.

Estimates of the information dimension for each time series at various points in time show some interesting attributes. For the Lorenz system, the estimate of the information dimension increased with time (Fig. 5.11, Table 5.1). The theoretical upper maximum for this dimension, the Lyapunov dimension (Farmer *et al.*, 1983; Parker & Chua, 1989), is approximately 2.01. The estimate of the information dimension for the Gaussian white noise time series also increased with time (Fig. 5.12, Table 5.1), approaching its theoretical upper maximum of 1. The Gaussian brown noise time series, however, had a remarkably constant value over time of around 0.92 (Fig. 5.12, Table 5.1).

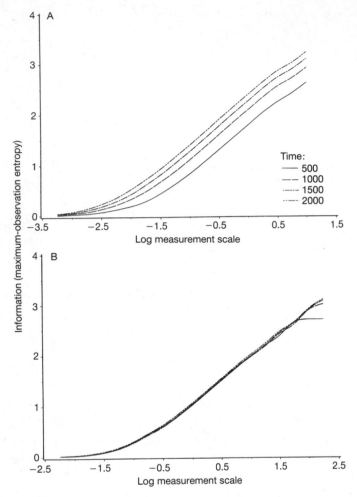

Fig. 5.12 Statistical information as a function of the logarithm of measurement scale (*h*) and time (different points in the time series are represented by different curves) for Gaussian white noise (A) and Gaussian brown noise (B). Note that there is no increase in statistical information over time for Gaussian brown noise.

Although these analyses should be considered preliminary, they do suggest that there are differences in the evolution of information between systems governed by deterministic chaos and those governed by constrained random processes. The increasing estimates of the information dimension associated with the Lorenz system are very different from the essentially constant information dimension of Gaussian brown noise. It is not clear how general these properties will be, and I suggest more work be done along these lines to establish theoretical guidelines for

Table 5.1 Estimates (standard errors in parentheses) of the slope of cumulative statistical information against the logarithm of measurement scale for several different time series at several points in the series

Time series	Cumulative portion of time period (quarters)			
	First	Second	Third	Fourth
Lorenz system	1.67 (0.04)	1.76 (0.02)	1.80 (0.02)	1.84 (0.01)
Gaussian white noise	0.92 (0.01)	0.95 (0.01)	0.96 (0.004)	0.97 (0.004)
Gaussian brown noise	0.92 (0.003)	0.92 (0.009)	0.92 (0.008)	0.93 (0.008)
American Redstart	0.80 (0.04)	0.81 (0.03)	0.82 (0.03)	0.82 (0.03)
Prairie Warbler	0.73 (0.05)	0.71 (0.06)	0.70 (0.06)	0.69 (0.06)

interpreting analyses of information dimensions for empirical time series.

5.3.3 Example analyses: dynamics of American Redstart and Prairie Warbler geographical populations

In this section I describe analyses of information dimensions of the time series generated by BBS censuses of two species of neotropical migratory wood warblers, American Redstart (*Setophaga ruticilla*) and Prairie Warbler (*Dendroica discolor*). Prairie Warbler is relatively narrowly distributed, being confined primarily to the southeastern United States (Fig. 5.13). This species occupies a relatively specialized habitat consisting primarily of early stages of succession and other open habitats. It has a semivariance dimension of approximately 2, indicating that there are steep changes in abundance from one place to the next in its geographical range. American Redstarts are distributed more widely across North America, extending across the northern part of the continent and down into the southeastern portion of the United States (Fig. 5.13). Redstarts are more generalized in their use of habitats, being found in a wider range of successional stages and vegetation types. The semivariance dimension is approximately 1.85, suggesting a more gradual degree of change in the species' abundance across space.

For each of these species, BBS abundances were used to construct probability distributions as follows. Each year, the abundances for the species on each route were accumulated for that year and for all previous years. A value for h, the measurement interval, was chosen to determine the abundance intervals, and the probability distribution was calculated by counting the number of BBS censuses that fell in

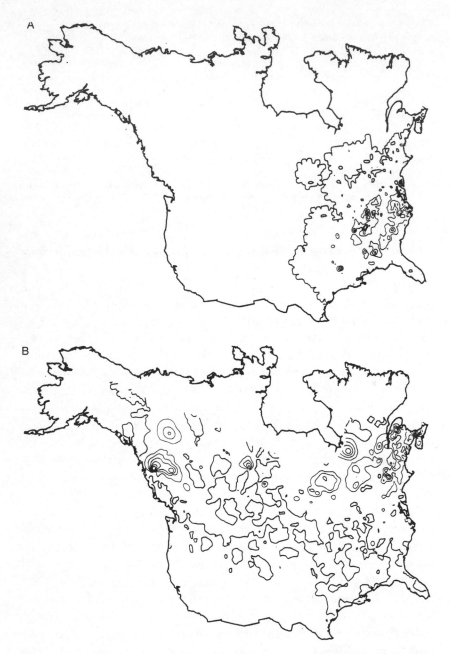

Fig. 5.13 Geographical of abundances for Prairie Warblers (A) and American Redstarts (B).

each abundance interval divided by the cumulative number of counts. The statistical entropy and information were calculated for the distribution using equations 5.3 and 5.4. This procedure was repeated for a dozen values of h. By the end of 23 years of census data, about 9200 observations had been accumulated for American Redstart and 5900 for Prairie Warbler. Statistical information values were then plotted against the logarithm of the measurement scale to determine the information dimension of the time series for each species.

The plots of statistical information against log measurement scale were approximately linear and the slope of the plots for Prairie Warbler appeared to be shallower than for American Redstarts (Fig. 5.14). The information dimension calculated from these plots did not change substantially over time for either species. Information dimensions were considerably less than 1 for both species, but were lower for the more narrowly distributed Prairie Warbler than for the American Redstart.

These results suggest several interesting possibilities. Firstly, both time series seem to have the same type of temporal dynamics of information as the Gaussian brown noise time series. That is, they seem to be indicative of some kind of constrained random motion. This would suggest that the population dynamics of both species have a substantial random component, but that randomness is constrained. Secondly, the Prairie Warbler seems to have a higher degree of constraint, since the estimates of the information dimension for this species were consistently lower than for the Redstart. This accords well with the observation that the Prairie Warbler is more narrowly distributed than the Redstart. Finally, the time series for both species differed in one significant way from the Gaussian brown noise time series. In that time series, statistical information did not increase over time, but remained roughly the same after a brief initial increase (Fig. 5.12). For both species, statistical information continued to increase over time. Thus, in some ways, the time series for both species were similar to the brown noise time series; in other ways they were more similar to a deterministic nonlinear process.

Although more needs to be done with the information dimension of time series, these preliminary results seem to suggest that for the two species studied, elements of both constrained random motion and deterministic chaos might be present. This opens up the intriguing possibility that there might be some fundamental relationship between deterministic chaos and constrained random motion. The dynamics of local populations of birds are determined by the demographics of individuals attempting to undergo their life histories, and there are an

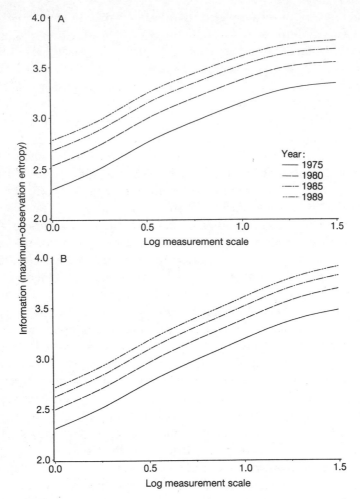

Fig. 5.14 Statistical information as a function of the logarithm of measurement scale (h) and time (different points in the time series are represented by different curves) for the geographical population dynamics of Prairie Warblers (A) and American Redstarts (B). The curves do not converge at small measurement scales because the values for the population abundance of each species are discrete (the smallest possible value being a single individual).

extremely large number of possible factors that influence these demographical processes. These demographical processes operate over large areas to result in the dynamics of a geographical population. The deterministic components of these processes may be sufficiently influential to act as strong constraints on the random factors affecting geographical patterns of population change in a species, so that the resulting time series appears to have both deterministic and stochastic components.

5.4 Summary

The analysis of temporal dynamics of geographical populations will be limited by the short period of time over which most of these populations must be observed. We have no good ideas regarding what time scales are appropriate to observe geographical populations; we can only be certain that they are longer than we have been able to observe geographical populations in the past. It is not unreasonable to expect that geographical populations change over scales of hundreds to thousands of years, and that even the longest systematic censuses are still too short to reveal much of the real dynamics of these populations.

Patterns of rates of change of geographical populations over space can be analysed by taking logarithms of spatial trend surfaces for adjacent time periods and subtracting them from one another, as suggested by Taylor (1986). Example analyses for two species with very different ecological and geographical range attributes show quite different results. The narrowly distributed neotropical migrant, Cerulean Warbler, had large changes in abundance concentrated in the centre of its geographical range. At the other extreme, Brown-headed Cowbird, a widespread species of brood parasite, showed no tendency to have large changes in abundance concentrated anywhere in its geographical range.

Analyses of the relationship between mean abundance and the standard deviation of abundance for each of these species seemed to substantiate this observation. There was also a significant positive relationship between spatial patterns of temporal variability and temporal patterns of spatial variability for a number of species. The existence of such patterns suggests that there is yet much to be learned from the examination of the temporal dynamics of geographical populations.

Perhaps the most interesting aspect of analysing the dynamics of geographical populations is the possibility that they may be nonlinear dynamical systems. Techniques for analysing time series generated by such processes have just recently been developed. There are two general classes of nonlinear dynamical systems that might be used as models of geographical populations. The first is that they are low-dimensional chaotic systems, that is, that their dynamics are determined by a relatively small number of nonlinear differential equations. A second type of model is that of a constrained random process. In such a process, the statistical likelihood of events varies among events: some events are much more likely than others because the past history of the system constrains it to a relatively small subset of possible behaviours. Examination of time series for the geographical populations of two species of wood warblers suggests that geographical populations may behave more like constrained random processes than chaotic processes. Clearly, much more work needs to be done before it will be possible to make any clear distinctions.

Understanding how geographical populations change over time is a critical piece of information that is missing in most discussions of the conservation of biological diversity. A number of questions arise that we just do not seem to have good answers for at the moment. Are there particular temporal patterns in geographical population change that indicate whether one species is more likely to become extinct than another? If so, we could begin to develop objective ways of identifying species that are rapidly approaching extinction. For example, if the information dimension can be interpreted as a measure of the population resilience of a species, then species with low resilience could be identified and conservation plans developed for them. In addition to such practical benefits, a better understanding of geographical population dynamics could provide a clearer insight into the basic processes involved in the evolution of biological diversity by helping us to understand phenomena like speciation and extinction. There is ample justification for more research in this area.

CHAPTER 6

The challenges of geographical population analysis

In the preceding pages, I have described some of the more interesting techniques that may be useful for the study of geographical populations. In this chapter I outline what I perceive to be some of the major questions for which the techniques developed herein might be useful in providing answers. I should note that the ideas that I discuss in this chapter are not necessarily a complete list of those that could be addressed using the techniques discussed in this book. I only hope to give the reader a flavour for the kinds of things that might be done.

6.1 Geographical range fragmentation and biodiversity

One problem facing conservationists is the ability to identify species that are likely to respond negatively to human-induced changes in the global ecosystem. It is becoming clearer to many that ecological processes in local communities are influenced by processes occurring at much larger scales than the local plots traditionally studied to elucidate them (Ricklefs, 1987; Brown & Maurer, 1989; Menge & Olson, 1990). These large-scale processes must affect local populations across relatively large spatial scales. To understand why some species may be more sensitive to global change than others, it is necessary to study geographical populations of species rather than to focus on individual local study plots.

It is difficult, or even impossible, to study the time course of the extinction of a species within the lifetime of an individual scientist. Even spectacular extinction events, such as the dramatic decline of Passenger Pigeons in North America, take decades to occur. Most continent-wide monitoring programmes have been in existence no longer than several decades. Some studies of local populations have been in place for a slightly longer period of time, but these are often of limited geographical extent. In the face of these spatial and temporal limitations on sampling, it might still be possible to begin to understand the dynamics of extinction by studying spatial patterns of population variation and comparing these among many species.

The basic approach would be to compare the current spatial patterns of geographical range fragmentation among species. By using sets of closely related species that share similar sets of ecological characteristics, such comparative analyses might provide a series of 'snapshots' of species in different stages of extinction (i.e. from widespread species that have reached the maximum size of their geographical range to narrowly distributed species that are likely to decline significantly). By examining characteristics of the geographical ranges of such series of species, it might be possible to form some idea of the time course of a single species as it progresses towards extinction. I have presented a number of such comparative analyses in the preceding pages, but have not examined their implications to any degree. My intent in this book has been to illustrate the methods rather than to explain the resulting patterns.

There are several difficulties with the comparative approach I have described above. It is possible that the patterns in collections of snapshots of individual species may be correlated with phylogeny. It is important to study the relationships between these patterns and phylogenetic relationships among species because they may give us some clues as to how to interpret the patterns in an evolutionary context. By knowing the relationships among species, we may be able to see relationships between the position of a species in its clade, and its geographical range characteristics. For example, are the oldest species in clades (i.e. those that are located near the 'root' of a cladogram) more or less likely to have highly fragmented geographical ranges, or is fragmentation unrelated to phylogenetic position? Answers to such questions would provide important information about the interpretation of the patterns seen among species.

Unfortunately, phylogenetic analyses are rarely available for large groups of species. For the North American birds that I have studied, the best 'phylogenies' available are either phenetic analyses of a few species within a higher taxon (e.g. Sibley & Ahlquist, 1990), or very restricted analyses of a few species. The situation is similar in many other taxa. Until more reliable studies are available that resolve both species- and higher-level phylogenies for a relatively large number of taxa, we will be extremely limited in what can be inferred about the relationships between phylogeny and geographical range fragmentation.

A second problem with the comparative approach is that some species that are currently narrowly distributed may have always been restricted in their geographical range whereas others may have been formerly widespread, but now have declined as continents have changed biotically since their origin. It is abundantly clear that local biological assemblages have changed dramatically since the Pleistocene (e.g. Graham, 1986) and entire biomes have changed dramatically as well

(Delcourt & Delcourt, 1991). In the face of such environmental change, it is unlikely that any species' geographical range has remained unaffected. But there is the possibility that some species that are currently relatively restricted in their abundance and distribution may have been formerly widespread, while other species may have always been relatively restricted. The comparative approach would use both species to represent a later stage of a species' progression towards extinction.

The use of the comparative method, however, can provide in the absence of temporal data some approximations of what we might expect of species that are progressing towards extinction or of species that are likely to respond negatively to widespread ecological change. I think that such data could be used to develop strategic approaches to biodiversity. Here I use 'strategic' in the sense of Gore (1992). Current conservation approaches to biological diversity are mainly 'tactical', that is, they develop specific plans for individual species in small geographical regions where the species' local population is threatened with local extinction. In some cases, such as the California Condor, this approach is equivalent to saving the entire species since the species has only a single population left. But there are also strategic threats to biodiversity, as Gore points out. Such threats include the potential for widespread decline of many species simultaneously over their entire geographical ranges. One such strategic threat has been the widespread decline of birds that breed in the deciduous forest of eastern North America and migrate to the neotropics during the winter (Robbins *et al.*, 1989). Conservation approaches for such strategic challenges must be more than tactical responses to declines of individual species in a few places. We must be able to prioritize species for tactical planning and to monitor the responses of the species to implementations of specific conservation efforts. Both of these long-term strategic goals can be implemented by using some of the techniques of geographical population analysis in combination with comparative analyses.

6.2 Patterns for organisms other than birds

All of the specific analyses reported here have used data on North American birds. This has been by design. This group of birds is one of the few that have had regular, extensive censuses conducted for a relatively long period of time (25 years) over a large geographical area (North America above the Mexican−United States boundary). There are other data sets available, notably the bird and insect censuses of Great Britain and the bird atlas of Australia. But these data sets have limitations that the BBS does not, so I focused my illustrations of the methods discussed in this book on the BBS. It is not unreasonable to ask what patterns might be evident if other kinds of organisms were used.

Birds are biologically unique in that they are highly vagile. Dispersal of individuals from their natal sites tends to be leptokurtic, however individuals can move relatively large distances to escape poor ecological conditions. Furthermore, some species show regular patterns of migration over long distances between breeding seasons, and there is probably a great deal of shuffling of individuals that goes on from one breeding season to the next. Because of these characteristics, it is unclear how generalizable the results obtained from studies of birds will be to other kinds of organisms. Virtually all plants are sedentary during most of their life history, and many insects and other specialized herbivores must remain localized in order to keep in contact with their hosts. Although these very different ecological circumstances will most likely lead to very different kinds of geographical ranges, it would be interesting to see if there is any systematic variation in geographical range parameters among taxa with different ecologies. Such systematic variation may actually be able to elucidate the kinds of ecological characteristics that are important in determining the different kinds of properties that geographical population analysis attempts to identify. Clearly, there is a great need for both better geographical censuses and for more analyses of many different kinds of organisms.

6.3 Macroecology and comparative biogeography

With the realization that there are more processes that occur at larger scales than local communities that might be important to the functioning of ecological systems, Brown and Maurer (1989) suggested that comparisons of large collections of related species might reveal the kinds of ecological processes that occur at the larger spatial scales. They identified patterns in the distribution of bird and mammal species among variables such as body mass and geographical range size that implied the operation of large-scale constraints on the geographical distribution and ecological attributes of species. Insights from such studies provide hypotheses that can be tested with data from different kinds of organisms to examine the generality of the patterns and processes.

Studies of geographical range fragmentation may provide yet another set of properties that are related to processes affecting assemblages of species on continents. For example, Brown (1984) hypothesized that many geographical ranges should have a single centre of abundance, and that abundance should decline relatively smoothly away from that centre. His model for this pattern incorporated both geographical variation in the environment and variation in the ecological requirements of species (see also Maurer & Brown, 1989). His model could be extended to include the possibility of more than one centre of abundance, but he emphasized that he thought in most instances, species would have only one centre of abundance. Examination of maps of abundance distribution of wintering (Root, 1988b) and breeding (unpublished maps in our laboratory) birds in North America indicates that there are relatively few species with what could be considered a single abundance peak. Maurer and Villard (in press) suggested that even in the absence of a majority of single-peaked abundance distributions, patterns of geographical range fragmentation could be interpreted as resulting from the same kinds of demographical process that Brown envisioned.

Understanding the behaviour of large scale ecological systems is a prerequisite to developing a better understanding of how continental and global processes result in ecological change. Therefore, macroecology has the potential to provide insights into how managers might approach large-scale systems. Clearly, many assumptions underlying many conventional approaches to managing ecosystems are not met in real ecosystems (e.g. Botkin, 1990; Pickett *et al.*, 1992), and the misapplication of them to management policy and decision-making can lead to poor management of ecological resources. The expanded perspective of macroecology can provide part of the necessary theoretical foundation for scaling up conservation decisions.

6.4 Fractals and chaos as models for biogeographical patterns

One of the most intriguing developments in modern mathematics is the discovery of nonlinear dynamics and fractal geometry. These two branches of mathematics are related in some very fundamental ways, and both have been suggested as potential modelling paradigms for many natural systems. Areas where potential applications exist include climatology, geomorphology and population dynamics. Some controversy has existed in the latter application, since most biological populations have not been studied for long enough to provide sufficient data for use in determining whether or not biological populations follow 'chaotic' dynamical laws.

Geographical populations, however, provide ideal systems for analysis using fractal geometry and nonlinear dynamics. The processes that affect them clearly operate simultaneously at numerous scales leading to the expectation of at least statistical self-similarity existing within them. The measures of geographical range shape and complexity, of spatial variation in population abundance and of geographical population dynamics show intriguing patterns in the comparative analyses reported here.

It should be noted, however, that the use of these novel mathematical techniques to analyse geographical populations in this book has been phenomenological. At present, we have no good theoretical population models that predict that the spatial patterns exhibited by species across their geographical ranges should necessarily be fractals. We have simply reasoned that in general, the kinds of processes that affect the distribution and abundance of individual species should lead us to expect fractal patterns. This, however, is not the same as saying that from first principles, we can deduce that geographical populations should have fractal patterns. Clearly, more theoretical work needs to be done on the consequences of spatial variation in population processes as they interact with spatial variation in ecological resources. Both mathematicians and theoretical ecologists could benefit by working on this problem in more detail.

6.5 Summary and conclusions

In this book I have outlined an approach to the analysis of entire populations of species across their geographical ranges that makes the basic assumption that these populations are relatively discrete entities that interact with ecological conditions in space to produce complicated patterns of abundance and distribution. The results of many of the comparative analyses that I have presented suggest that this perspective may prove useful in developing a better understanding of how species use geographical space. This in turn may provide a basis for developing strategic approaches to the management of biological diversity.

The size and shape of geographical ranges are important in understanding the nature of the boundaries of species' geographical ranges. Clearly, such boundaries are diffuse, and change with the scale at which we observe them. However, the rate at which these boundaries change as the scale of observation changes may be a constant that is unique to each species. This is the idea behind using the box dimension as a measure of range boundary complexity.

The nature of how abundance changes from one place in one species' geographical range to the next provides additional insight into the nature of geographical populations of species. The distribution of population abundance across space can be estimated using a number of techniques derived from the geostatistical literature (although some of these techniques were borrowed by geostatisticians from more basic research in statistics). Since species are distributed patchily at all scales, from individuals within local habitats, to populations among landscapes, to regions within entire continents, the use of measures that describe how abundance changes with changing scale is desirable. Examination of analyses of individual species implies that there are not single scales of variation in abundance, but that there may be multiple scales of variation, each potentially related to different processes.

Changes in geographical populations over time can be studied, albeit with great difficulty, by long-term censuses of entire geographical populations. The resulting patterns obtained by analysis of such data tend to be exceedingly complex. However, there is the suggestion that narrowly distributed species tend to have most of the large changes in their populations associated with regions of high abundance, while for more widely distributed species, large changes are diffused throughout the geographical range. When examining how abundances change over time, the use of statistical entropy and information provide a way of assessing the degree of complexity in those dynamics. Geographical populations seem to change more like constrained, random systems than strictly chaotic systems, but the time-scales over which many have been observed certainly should lead one to accept such a conclusion as provisional at best.

Geographical population analysis has the potential to lead to new insights in strategic planning for the conservation of biological diversity. By providing basic analyses on how species are distributed across space and react to continental environments, the techniques discussed in this book allow comparisons among species that might help us understand what kind of species tend to be more threatened by massive changes in global ecosystems. When used in conjunction with long-term monitoring programs, they may provide a basis for evaluating both successes and failures of strategic management decisions across entire geographical ranges of species.

The tools and basic data for studying geographical populations are few, and useable data are available for only a handful of taxa. Perhaps as techniques for the remote sensing of ecosystems become more sophisticated, it may become possible to obtain large scale data sets for taxa other than birds. Such data is critical, in my mind, if we are to begin to develop a complete picture of how we must manage ecological resources across large scales. Human activities have led to massive changes in ecological systems across virtually the entire earth. We must develop the ability to describe, understand and react to these changes, indeed, the survival of our species may depend on it. It is my hope that geographical population analysis will provide a piece of the puzzle, albeit a very important one. The techniques I have described herein represent just a few of the many that may become available in the future as computing and remote sensing technologies evolve. Ecologists, biogeographers and conservationists should be prepared to take advantage of these new tools as they develop.

References

Allen T.F.H. & Starr T.B. (1982) *Hierarchy*, 310 pp. University of Chicago Press, Chicago [1.1.1]

Barnsley M. (1988) *Fractals Everywhere*, 394 pp. Academic Press, Boston [3.2.1, 3.2.2]

Begon M., Harper J.L. & Townsend C.R. (1990) *Ecology*, 945 pp. Blackwell Scientific Publications, Cambridge, MA [1.1]

Botkin D.B. (1990) *Discordant Harmonies*. Oxford University Press, Oxford [6.3]

Brooks D.R. & McLennan D.H. (1991) *Phylogeny, Ecology, and Behavior*, 434 pp. University of Chicago Press, Chicago [1.1.2]

Brooks D.R. & Wiley E.O. (1988) *Evolution as Entropy*, 2nd edn. 415 pp. University of Chicago Press, Chicago [5.3.2]

Brown J.H. (1984) On the relationship between abundance and distribution of species. *American Naturalist*, **124**, 225−279 [1.2.1, 1.2.2, 4.1.2, 6.3]

Brown J.H. & Maurer B.A. (1987) Evolution of species assemblages: effects of energetic constraints and species dynamics on the diversification of the North American avifauna. *American Naturalist*, **130**, 1−17 [3.1.2, 3.1.3]

Brown J.H. & Maurer B.A. (1989) Macroecology: the division of food and space among species on continents. *Science*, **243**, 1145−1150 [3.1.2, 6.1, 6.3]

Burgess T.M. & Webster R. (1980a) Optimal interpolation and isarithmic mapping. I. The semivariogram and punctual kriging. *Journal of Soil Science*, **31**, 315−332 [4.1.1]

Burgess T.M. & Webster R. (1980b) Optimal interpolation and isarithmic mapping. II. Block kriging. *Journal of Soil Science*, **31**, 333−342 [4.1.1]

Burrough P.A. (1983) Multiscale sources of spatial variation in soil. I. The application of fractal concepts to nested levels of soil variation. *Journal of Soil Science*, **34**, 577−597 [4.2.1]

Burrough P.A. (1986) *Principles of Geographical Information Systems for Land Resources Assessment*, 194 pp. Oxford University Press, Oxford [2.1, 2.1.1, 4.1.1, 4.2.1]

Cressie N. (1991) *Statistics for Spatial Data*, 900 pp. Wiley, New York [4.1.1]

Davis J.C. (1986) *Statistics and Data Analysis in Geology*, 2nd edn., 646 pp. Wiley, New York [4.1.1]

Delcourt H.A. & Delcourt P.A. (1991) *Quaternary Ecology*, 242 pp. Chapman & Hall, London [6.1]

Edelstein-Keshet L. (1988) *Mathematical Models in Biology*, 586 pp. Random House, New York [2.1]

Ehrlich P.R. & Roughgarden J. (1987) *The Science of Ecology*, 710 pp. MacMillan, New York [1.1]

Eldredge N. (1985) *Unfinished Synthesis*, 237 pp. Oxford University Press, New York [1.1.1]

Emlen J.T., DeLong M.J., Jaeger M.J., Moermond T.C., Rusterholz K.A. & White R.P. (1986) Density trends and range boundary constraints of forest birds along a latitudinal gradient. *Auk*, **103**, 791−803 [4.1.3]

Erwin T.L. (1991) How many species are there?: revisited. *Conservation Biology*, **5**, 330−333 [1.1.2]

Falconer K. (1990) *Fractal Geometry*, 288 pp. Wiley, New York [3.2.1]

Farmer J.D., Ott E. & Yorke J.A. (1983) The dimension of chaotic attractors. *Physica* **7D**, 153−180 [5.3.2]

Ford H.A. (1990) Relationships between distribution, abundance, and foraging specialization in Australian landbirds. *Ornis Scandinavica*, **21**, 133−138 [3.1.2]

Frankel O.H. & Soulé M.E. (1980) *Conservation and Evolution*. Cambridge University Press, Cambridge [1.1.2]

Gaston K.J. (1991a) The magnitude of global insect species' richness. *Conservation Biology*, **5**, 283−296 [1.1.2]

Gaston K.J. (1991b) Estimates of the near-imponderable: a reply to Erwin. *Conservation Biology*, **5**, 564−566 [1.1.2]

Gaston K.J. (1991c) How large is a species' geographic range? *Oikos*, **61**, 434−438 [3, 3.1.3]

Gilpin M.E. (1988) A comment on Quinn and Hastings: extinction in subdivided habitats. *Conservation Biology*, **2**, 290−292 [1.2]

Gilpin M. & Hanski I. (eds.) (1991) *Metapopulation Dynamics: Empirical and Theoretical Investigations*, 336 pp. Academic Press, New York [1.2, 1.2.2]

Goodman D. (1987a) Consideration of stochastic demography in the design and management of biological reserves. *Natural Resource Modeling*, **1**, 204−234 [5.2]

Goodman D. (1987b) The demography of chance extinction. In Soulé H.E. (ed.) *Viable Populations for Conservation*, pp 11−34. Cambridge University Press, Cambridge [5.2]

Gore A. (1992) *Earth in the Balance*. Penguin Books, New York [6.1]

Graham R.W. (1986) Response of mammalian communities to environmental changes during the Late Quaternary. In Diamond J. & Case T.J. (eds) *Community Ecology*, pp 300−313. Harper & Row, New York [6.1]

Gst (1993) *Geostatistics for the Environmental Sciences*. Gamma Design Software, Plainwell, Michigan [4.1.1, 4.1.2]

Haining R. (1990) *Spatial Data Analysis for the Social and Environmental Sciences*, 409 pp. Cambridge University Press, Cambridge [2.1, 2.1.2, 4.1.1, 4.2.1]

Hanski I. (1982) On patterns of temporal and spatial variation in animal populations. *Annales Zoologici Fennici*, **19**, 21−37 [5.2.1]

Hengeveld R. (1990) *Dynamic Biogeography*, 249 pp. Cambridge University Press, Cambridge [1.2.1, 3, 3.1, 4.1.2, 5]

Hengeveld R. & Haeck J. (1981) The distribution of abundance. II. Models and implications. *Proceedings of the Koninklijke Nederlandse Akademie van Wetenshappen, C*, **84**, 257−284 [1.2.1, 3.1]

Hengeveld R. & Haeck J. (1982) The distribution of abundance. I. Measurements. *Journal of Biogeography*, **9**, 303−316 [1.2.1, 1.2.2]

Heywood S.G. 'Fractal analysis of spatial patterns of abundance using Breeding Bird Survey data'. (Master's Thesis, Brigham Young University, Provo, Utah, 1992), 92 pp. [5.1.1]

Jandel (1992) *TableCurve Automated Curve Fitting Software*, 262 pp. Jandel Scientific, San Rafael, California [4.1.3]

Kendeigh S.C. (1944) Measurement of bird populations. *Ecological Monographs*, **14**, 67−106 [4.1.3]

Kineman J.J. & Ohrenschall M.A. (1992) Global ecosystems database. US Department of Commerce, NOAA, National Geophysical Data Center, Boulder, CO [4.1.4]

Kolasa J. (1989) Ecological systems in hierarchical perspective: breaks in community structure and other consequences. *Ecology*, **70**, 36−47 [1.2.1]

Lande R. (1988) Genetics and demography in biological conservation. *Science*, **241**, 1455−1460 [1.1.2, 1.2.2]

Lasota A. & Mackey M.C. (1994) *Chaos, Fractals and Noise: Stochastic aspects of dynamics*. Springer-Verlag, New York [5.3.2]

Lawton J.H. (1989) What is the relationship between population density and body size in animals? *Oikos*, **55**, 429−434 [3.1.2]

Legendre P. & Fortin M.-J. (1989) Spatial patterns and ecological analysis. *Vegetatio*, **80**, 107−138 [4]

Leigh E.G. Jr. (1981) The average lifetime of a population in a varying environment. *Journal of Theoretical Biology*, **90**, 213−239 [5.2]

Lorenz E.N. (1963) Deterministic nonperiodic flow. *Journal of Atmospheric Science*, **20**, 130−141 [5.3.1]

Lotka A.J. (1925) *Elements of Physical Biology* (1956 Dover reprint), 465 pp. Dover, New York [5.3.1]

Lovejoy S. (1982) Area−perimeter relation for rain and cloud areas. *Science*, **216**, 185−187 [3.2.2]

MacArthur R.H. (1972) *Geographical Ecology*. Harper & Row, New York [5.3.1]

Mandelbrot B.B. (1983) *The Fractal Geometry of Nature*, 468 pp. Freeman, New York [2.1.3, 3.2.1]

Matheron G. (1963) Principles of geostatistics. *Economic Geology*, **58**, 1246−1266 [2.1]

Maurer B.A. (1993) Evolution of information in nonlinear systems. In Collier J. & Siegel-Causey D. (eds) *A Unified Approach to Biology: Recent Work on Non-Equilibrium Macrosystems*, in press [5.3.2]

Maurer B.A. & Brown J.H. (1988) Distribution of energy use and biomass among species of North American terrestrial birds. *Ecology*, **69**, 1923−1932 [3.1.2]

Maurer B.A. & Brown J.H. (1989) Distributional consequences of geographic variation in demographic processes. *Annales Zoologici Fennici*, **26**, 121−131 [1.2.2, 4.1.2, 6.3]

Maurer B.A., Ford H.A. & Rapoport E.H. (1991) Extinction rate, body size, and avifaunal diversity. *Acta XX Congressus Internationalis Ornithologici*, **2**, 826−834 [3.1.2]

Maurer B.A. & Heywood S.G. (1993) Geographic range fragmentation and abundance in neotropical migratory birds. *Conservation Biology*, **7**, 501−509 [1.2.1, 4.1.2, 4.2.1, 5.1.1]

Maurer B.A. & Villard M.-A. (1993) Geographic variation in abundance of North American birds. *Research & Exploration*, in press [1.2.1, 3.1.1, 4]

May R.M. (1988) How many species are there on earth? *Science*, **241**, 1441−1449 [1.1.2]

May R.M. (1990) How many species? *Philosophical Transactions of the Royal Society of London B*, **330**, 293−304 [1.1.2]

Menge B.A. & Olson A.M. (1990) Role of scale and environmental factors in regulation of community structure. *Trends in Ecology and Evolution*, **5**, 52−57 [6.1]

Milne B.T. (1991) Lessons from applying fractal models to landscape patterns. In Turner M.G. & Gardner R.H. (eds) *Quantitative Methods in Landscape Ecology*, pp 199−235. Springer-Verlag. Berlin [2.1.1, 3.2.2]

Morse D.R., Stork N.E. & Lawton J.H. (1988) Species number, species abundance and body length relationships of arboreal beetles in Bornean lowland rain forest trees. *Ecological Entomology*, **13**, 25−37 [3.1.2]

Norse E.A., Rosenbaum K.L., Wilcove D.S., Wilcox B.A., Romme W.H., Johnston D.W. & Stout M.L. (1986) *Conserving Biological Diversity in our National Forests*, 116 pp. The Wilderness Society [1.1]

Office of Technology Assessment (1987) Technologies to maintain biological diversity. *OTA-F*-330, US Congress, Washington, DC [1.1]

O'Neill R.V., DeAngelis D.L., Waide J.B. & Allen T.F.H. (1986) *A Hierarchical Concept of Ecosystems*, 253 pp. Princeton University Press, Princeton, NJ [1.1.1]

Ord J.K. (1979) Time series and spatial patterns in ecology. In Cormack R.M. & Ord J.K. (eds) *Spatial and Temporal Analysis in Ecology*, pp 1−94. International Cooperative Publishing, Fairland, MD [4]

Otte D. & Endler J.A. (eds) (1989) *Speciation and its Consequences*, 679 pp. Sinauer, Sunderland, MA [1.1.2]

Palmer M.W. (1988) Fractal geometry: a tool for describing spatial patterns of plant communities. *Vegetatio*, **75**, 91–102 [2.1.3]

Parker T.S. & Chua L.O. (1989) *Practical Numerical Algorithms for Chaotic Systems*, 348 pp. Springer-Verlag, New York [3.2.2, 5.3.2]

Pearson K. (1901) On lines and planes of closest fit to systems of points in space. *Philosophical Magazine*, **2**, 559–572 [3.1.1]

Pennycuick C.J. & Kline N.C. (1986) Units of measure for fractal extent, applied to the coastal distribution of bald eagle nests in the Aleutian Islands, Alaska. *Oecologia*, **68**, 254–258 [4.2.1]

Peitgen H.O., Jurgens H. & Saupe D. (1992) *Chaos and Fractals: Frontiers of Science*. Springer-Verlag, New York [3.2.1]

Peitgen H.-O. & Saupe D. (eds) (1988) *The Science of Fractal Images*, 312 pp. Springer-Verlag, New York [3.2.1]

Pickett S.T.A., Parker V.T. & Fiedler P.L. (1992) The new paradigm in ecology: implications for conservation above the species level. In Fiedler P.L. & Jain S.K. (eds) *Conservation Biology*, pp 65–88. Chapman & Hall, New York [6.3]

Pielou E.C. (1977) *Mathematical Ecology*, 385 pp. Wiley, New York [3.1.1]

Pielou E.C. (1984) *The Interpretation of Ecological Data*, 263 pp. Wiley, New York, [3.1.1]

Pimm S.L. (1991) *The Balance of Nature?* 434 pp. University of Chicago Press, Chicago [1.3, 4.2.1, 5.2, 5.2.2, 5.3]

Pulliam H.R. (1988) Sources, sinks, and population regulation. *American Naturalist*, **132**, 652–661 [1.2]

Quinn J.R. & Hastings A. (1987) Extinction in subdivided habitats. *Conservation Biology*, **1**, 198–209 [1.2, 1.2.2]

Quinn J.R. & Hastings A. (1988) Extinction in subdivided habitats: a reply to Gilpin. *Conservation Biology*, **2**, 293–296 [1.2]

Ralph C.J. & Scott J.M. (eds) (1981) Estimating Numbers of Terrestrial Birds. *Studies in Avian Biology*, **6**, Cooper Ornithological Society Lawrence, KS [4.1.3]

Rapoport E. (1982) *Areography: Geographical Strategies of Species*. Pergamon Press, New York [1.2.1, 3]

Ricklefs R.E. (1987) Community diversity: relative roles of local and regional processes. *Science*, **235**, 167–171 [6.1]

Ricklefs R.E. (1990) *Ecology*, 3rd edn., 896 pp. Freeman, New York [1.1]

Ripley B.D. (1981) *Spatial Statistics*, 252 pp. Wiley, New York [2.1, 4.1.1]

Robbins C.S., Bystrak D. & Geissler P.H. (1986) *The Breeding Bird Survey: Its First Fifteen Years, 1965–1979*, 196 pp. Resource Publication 157, United States Department of the Interior Fish and Wildlife Service, Washington, DC [3.1.2]

Robbins C.S., Sauer J.R., Greenberg R.S. & Droege S. (1989) Population declines in North American birds that migrate to the neotropics. *Proceedings of the National Academy of Sciences USA*, **86**, 7658–7662 [6.1]

Robbins C.S., Fitzpatrick J.W. & Hamel P.B. (1992) A warbler in trouble: *Dendroica cerulea*. In Hagan J.M. III & Johnston D.W. (eds.) *Ecology and Conservation of Neotropical Migrant Landbirds* pp 549–562. Smithsonian Institution Press, Washington, DC [5.1.1]

Robinson S.K. (1992) Population dynamics of breeding Neotropical migrants in a fragmented Illinois landscape. In Hagan J.M. III & Johnston D.W. (eds) *Ecology and Conservation of Neotropical Migrant Landbirds*, pp 408–418. Smithsonian Institution Press, Washington, DC [5.1.1]

Root T. (1988a) Energy constraints on avian distributions and abundances. *Ecology*, **69**, 330–339 [3, 3.2.2]

Root T. (1988b) *Atlas of Wintering North American Birds*, 312 pp. University of Chicago Press, Chicago [4.1.1, 6.3]

Rossi R.E., Mulla D.J., Journel A.G. & Franz E.H. (1992) Geostatistical tools for modeling and interpreting ecological spatial dependence. *Ecological Monographs*, **62**, 277−314 [2.1.2]

Ruelle D. (1989) *Chaotic Evolution and Strange Attractors*, 96 pp. Cambridge University Press, Cambridge [5.3.2]

Salthe S.N. (1985) *Evolving Hierarchical Systems*, 343 pp. Columbia University Press, New York [1.1.1]

Sampson R.J. (1988) *Surface III*, 277 pp. Kansas Geological Survey, Lawrence, KS [4.1.1, 4.1.2]

Schaffer W.M. (1984) Stretching and folding in lynx fur returns: evidence for a strange attractor in nature? *American Naturalist*, **124**, 798−820 [5.3]

Schaffer W.M. (1985) Order and chaos in ecological systems. *Ecology*, **66**, 93−106 [5.3]

Schaffer W.M. & Kot M. (1985) Do strange attractors govern ecological systems? *BioScience*, **35**, 342−350 [5.3]

Schaffer W.M. & Kot M. (1986) Chaos in ecological systems: the coals that Newcastle forgot. *Trends in Ecology and Evolution*, **1**, 58−63 [5.3]

Schroeder M. (1991) *Fractals, Chaos, Power Laws: Minutes from an Infinite Paradise* 429 pp. Freeman, New York [3.2.1]

Sibley C.G. & Ahlquist J.E. (1990) *Phylogeny and Classification of Birds*, 976 pp. Yale University Press, New Haven, CT [6.1]

Skellam J.G. (1951) Random dispersal in theoretical populations. *Biometrika*, **38**, 196−218 [1.2, 1.2.2]

Smith R.L. (1990) *Ecology and Field Biology*, 4th edn., 922 pp. Harper & Row, New York [1.1]

Strayer D. (1986) The size structure of a lacustrine zoobenthic community. *Oecologia*, **69**, 513−519 [3.1.2]

Sugihara G. & May R.M. (1990) Applications of fractals in ecology. *Trends in Ecology and Evolution*, **5**, 79−86 [3.2.1]

Taylor L.R. (1961) Aggregation, variance and the mean. *Nature*, **189**, 732−735 [1.2, 5.2.1]

Taylor L.R. (1986) Synoptic dynamics, migration and the Rothamsted Insect Survey. *Journal of Animal Ecology*, **55**, 1−38 [1.2, 4.1]

Taylor L.R. & Taylor R.A.J. (1977) Aggregation, migration, and population mechanics. *Nature*, **265**, 415−421 [5.2.1]

Turner J.R.G., Gatehouse C.M. & Corey C.A. (1987) Does solar energy control organic diversity? Butterflies, moths, and the British climate. *Oikos*, **48**, 195−205 [1.1.1]

Turner J.R.G., Lennon J.J. & Lawrenson J.A. (1988) Species diversity: the seasonal distribution of British birds supports the energy-theory. *Nature*, **335**, 539−541 [1.1.1]

Voss R.F. (1988) Fractals in nature: from characterization to simulation. In Peitgen H.-O. & Saupe D. (eds) *The Science of Fractal Images* pp 21−70. Springer-Verlag, New York [3.2.2, 5.3.1]

Wiens J.A. (1989) *The Ecology of Bird Communities*, Vols 1 & 2, pp 539 & 316. Cambridge University Press, Cambridge [2]

Wiley E.O. (1981) *Phylogenetics*, 439 pp. Wiley, New York [1.1.1]

Williamson M.H. & Lawton J.H. (1991) Fractal geometry of habitats. In Bell S.S., McCoy E.O. & Mushinsky H.R. (eds) *Habitat structure: the physical arrangement of objects in space*, pp 69−86. Chapman & Hall, London [3.2.2]

Wright D.H., Currie D.J. & Maurer B.A. (1993) Energy supply and patterns of species richness on local and regional scales. In Ricklefs R.E. & Schluter D. (eds) *Species Diversity: Historical and Geographical Perspectives*, University of Chicago Press, Chicago [1.1.1]

Index